Optimization Design and Simulation Analysis of Hydraulic Turbine

Optimization Design and Simulation Analysis of Hydraulic Turbine

Guest Editors

Xiuli Mao
Yuchuan Wang

MDPI

Basel • Beijing • Wuhan • Barcelona • Belgrade • Novi Sad • Cluj • Manchester

Guest Editors
Xiuli Mao
College of Water Resources and Architectural Engineering
Northwest A&F University
Yangling
China

Yuchuan Wang
College of Water Resources and Architectural Engineering
Northwest A&F University
Yangling
China

Editorial Office
MDPI AG
Grosspeteranlage 5
4052 Basel, Switzerland

This is a reprint of the Special Issue, published open access by the journal *Energies* (ISSN 1996-1073), freely accessible at: https://www.mdpi.com/journal/energies/special_issues/505TR6KG1J.

For citation purposes, cite each article independently as indicated on the article page online and as indicated below:

Lastname, A.A.; Lastname, B.B. Article Title. *Journal Name* **Year**, *Volume Number*, Page Range.

ISBN 978-3-7258-5809-5 (Hbk)
ISBN 978-3-7258-5810-1 (PDF)
https://doi.org/10.3390/books978-3-7258-5810-1

Cover image courtesy of Yuchuan Wang

Contents

energies

MDPI

Article

Development of WHED Method to Study Operational Stability of Typical Transitions in a Hydropower Plant and a Pumped Storage Plant

Xiuli Mao [1,2,3,4,*], Guoqing Wen [1,4], Yuchuan Wang [1,4], Jiaren Hu [1,4], Xuetao Gan [1,4] and Pengju Zhong [1,4]

1 College of Water Resources and Architectural Engineering, Northwest A&F University, Yangling 712100, China; wenguoqing@nwafu.edu.cn (G.W.); wyc@nwafu.edu.cn (Y.W.); hjr@nwafu.edu.cn (J.H.); ganxuetao@nwsuaf.edu.cn (X.G.); cpj1998@nwafu.edu.cn (P.Z.)
2 Key Laboratory of Fluid and Power Machinery, Ministry of Education, Xihua University, Chengdu 610039, China
3 School of Fundamental Science and Engineering, Waseda University, Tokyo 169-8555, Japan
4 Key Laboratory of Agricultural Soil and Water Engineering in Arid and Semiarid Areas, Ministry of Education, Northwest A&F University, Yangling 712100, China
* Correspondence: maoxl@nwafu.edu.cn

Abstract: This study proposes the water hammer energy difference (WHED) method based on unsteady flow energy and continuity equations, as well as the propagation laws of water hammer in closed pipes, and verifies its accuracy. Additionally, the parameter evolution patterns of typical transient conditions in pumped storage power plants are investigated based on WHED. The application of WHED in the transient processes of hydropower plants (HPs) is validated by experiments, showing a maximum error of about 7% between numerical and experimental results under conditions of initial load increase followed by decrease (H_R = 184 m). Additionally, WHED was validated under two critical conditions in pumped storage plants (PSPs): 90% load rejection in generating mode and emergency power-off in pumping mode. In PSPs, the results of WHED are consistent with those obtained using the method of characteristics (MOC), with a maximum fault tolerance rate $\Delta < 3\%$. Notably, WHED offers superior time efficiency when analyzing hydraulic transitions in complex pipe networks, as it directly considers boundary conditions at both ends of the pipeline and hydraulic machinery, whereas MOC requires dividing the pipeline into multiple segments with a series of boundary points. Lastly, WHED's energy parameters are used to describe flow stability from a physics perspective, explaining the causes of pressure fluctuations during transient periods in HPs and PSPs. These findings offer valuable references and guidance for the safe operation of PSPs and HPs.

Keywords: hydropower plant; pumped storage plant; typical transient conditions; method of water hammer energy difference; method of characteristics

1. Introduction

The development of clean energy is a crucial approach to mitigating the global energy crisis faced by all countries both now and in the future [1,2]. As shown in Figure 1, the main use of clean energy is for power generation, with pumped storage plants (PSPs) significantly contributing to grid instability by compensating for the instability caused by renewable energy sources like photovoltaic and wind energy [3,4]. PSPs meet grid requirements by adapting their operation conditions, further aiding in the construction of a clean and smart grid [5,6]. In addition, to meet the dynamic demands of the power grid, more hydropower plants (HPs) are gradually transitioning from base load to load regulation roles. Regulating HPs and PSPs involves long-term and frequent operation under transitional conditions [7–9]. For example, the installed capacity of the PSP case in this study is 1.2 million kW (according to the plan of the National Energy Administration of China, by the end of 2025, the installed capacity of pumped storage power plants in China is expected to reach 62 million kW. Based on this figure, the case study power plant accounts for approximately 1.9% of the total installed capacity). Its annual auxiliary grid acceptance of new energy power is more than 200 million kW·h, and this includes an annual average of more than 1500 unit startups and shutdowns [10,11]. Consequently, transitions of regulated plants directly determine the success of the new energy grid connection and the stability of the grid operation, and failure can even lead to paralysis of the entire new energy grid system.

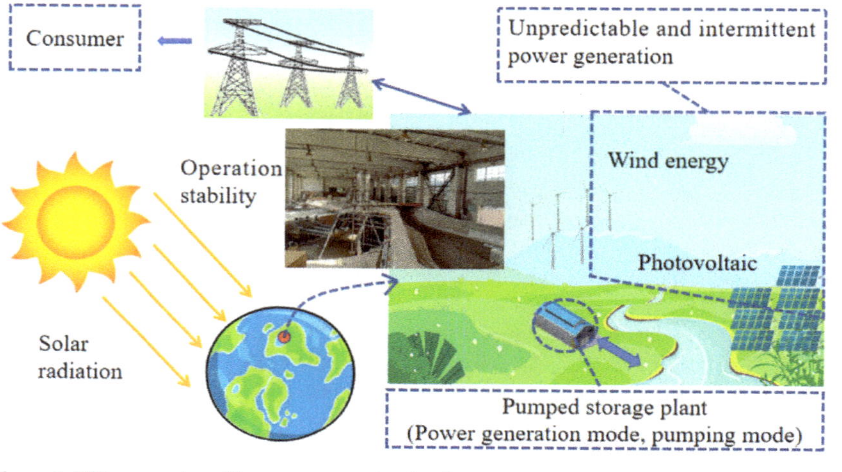

Figure 1. PSP cooperates with power generation by clean energy.

Transitions of PSPs include about 24 situations, such as transitioning from stationary to power generation or pumping with a full load, from a no load condition to full load in generating mode, and from phase modulation to a full load in pumping mode, among others [12,13]. Switching between generating and pumping modes with a full load involves experiencing pumping, braking, generating, and reverse pumping zones. These transient periods not only have significant impact on the unit, but the active power also increased by a factor of two compared to the rated condition in a short period [14,15]. Almost all operating PSPs and HPs today encounter dynamic problems, such as exceeding vibration and noise parameters, and even failures or accidents of units. Therefore, the operation stability of PSPs and HPs is a crucial issue studied by scholars in the field of hydroelectricity [16,17].

The water hammer phenomenon is a significant topic in fluid mechanics, with its discovery and research dating back to the 19th century. Scientists have conducted both theoretical studies and experimental validations to clarify the propagation characteristics of water hammer, with mitigating its adverse effects becoming a key research focus [18]. With advancements in computational technology, numerical simulation has become an essential approach for studying water hammer. In addition to using computational fluid dynamics (CFD) to predict water hammer effects, some researchers have proposed a simplified convolution integral method, which significantly improves the computational efficiency of water hammer simulations while maintaining high accuracy [19]. The public literature shows accidents caused by high water hammer pressure in many hydropower plants, which cause huge economic losses and human casualties [20,21]. The safety and stability of hydropower plants are key issues studied by scholars around the world, wherein the guarantee calculation of regulation in the pre-scientific phase is necessary for all plant engineering [22,23]. Suitable adjusting cases include the number of operating units, moving laws of guide vanes and valves, different water levels in the upper reservoir, etc., having significant influence on the stability and safety of the system, like the maximums of speed and pressure.

Effective engineering methods applied to improve the operating stability of PSPs and HPs include optimizing the guide vane closure law and placing the surge tank in a suitable position [24–26]. Reasonable flow control laws can significantly reduce pressure fluctuation. For example, pressure fluctuating amplitudes can be reduced by 25–35% if guide vanes are closed before the unit exits the grid, compared to when the unit is disconnected from the grid while the guide vanes remain open [27,28]. Co-adjustment of the guide vane and valves is another way to improve fluid stability, especially during the load rejection process of PSPs [26,29]. Aside from emergencies, the initial conditions of PSP and HP transitions are crucial for transient operation stability. Thus, regular switching of PSP and HP conditions can help adjust parameters of the initial condition first [30]. The above research pertains to the constructed and operating PSPs and HPs. However, during the pre-feasibility stage of PSPs and HPs, how can we explain the fluid stability in the flow passage? Consequently, developing a method to investigate transient conditions of PSPs and HPs with higher precision and better timeliness is the main objective of this work.

The one-dimensional (1D) linear method, three-dimensional (3D) numerical simulation, and combinations of 1D and 3D methods in different flow sections are the main approaches to investigate the transient characteristics of PSPs and HPs. The 1D linear method, based on finite differences, is widely used to calculate runner speed and mass-flow of PSPs and HPs. The traditional linear method is unable to characterize the evolution of water hammer from a physical point of view, and it is also computationally inefficient, whereas WHED can overcomes these limitations. Three-dimensional numerical simulation, on the other hand, provides full visualization and can investigate the external properties of structures [31,32]. However, 3D numerical simulation requires substantial computational resources, especially for fine computations, and it is not suitable for the pre-feasibility stage since only the hydraulic structures are designed, not the unit structure. For constructed and operating PSPs or HPs, some scholars adopt the 3D numerical method for the unit and the 1D approach for other sections like the water diversion system (WDS). This approach saves computational resources but reduces study precision due to variability between 1D and 3D data [33,34]. Some published works optimize the computational process of transitions in PSPs using partial overlap algorithms, reducing errors caused from the exchange of 1D and 3D results, but increasing operational difficulty and requiring more computational resources [35,36].

This work proposes a method of water hammer energy difference (WHED) to analyze the transient characteristics of PSPs and HPs under various conditions. The principles and derivation of WHED are introduced in Section 2. Section 3 presents the validation of WHED through two critical conditions of PSPs, with comparisons to physical model experiments. Section 4 discusses the application of WHED in the regulation insurance project of a PSP in China. Section 5 provides conclusions.

2. Method of Water Hammer Energy Difference (WHED)

2.1. Fundamental Theory of WHED

The WHED simulates the generation, propagation, transmission, reflection, and attenuation of water hammer in a closed pipeline, based on the energy and continuity equations for unsteady flow. Transient parameters are calculated by analyzing the energy of the water hammer at each time step. Assuming that the pipe wall is an incompressible rigid body and the water inside the closed pipe is an ideal fluid, the fluid satisfies the equations of motion and continuity. Water hammer in pipes is essentially the transmission of pressure waves [37]. The fluid particle at any moment can be described by Equation (1). The energy of transient fluid at a certain moment consists of potential energy $\frac{dH}{2}$, kinetic energy $\frac{adv}{2g}$, and loss term $\frac{\lambda dl}{2d}\frac{v|v|}{2g}$.

$$dH \pm \frac{a}{g}dv + \frac{\lambda v|v|}{2gd}dl = 0 \tag{1}$$

The initial water head and flow velocity at the center of the pipe are defined as H_0 and v_0. H_0 and v_0 are substituted into Equation (1), making a difference with the current moment (H and v) results in Equation (2), which represents the energy change of a fluid particle in the pipeline. Equation (2) is the basic expression of WHED.

$$\frac{d}{dl}\left(\frac{H-H_0}{2} \pm \frac{a}{2g}(v-v_0)\right) = -\frac{\lambda}{4gd}(v|v| - v_0|v_0|) \tag{2}$$

Figure 2 shows a random fluid particle P in the pipeline, flanked by points B ($L_B = z - z_B$) and C ($L_C = z_C - z$), where the water flows from B to C. Integrating the length l in Equation (2) along the pipe yields Equations (3) and (4). Assuming the valve at B is closed, a positive wave in the same direction as the incoming flow appears to the right of B and propagates towards C. When this positive wave arrives at C, a negative wave is produced due to reflection, which propagates towards B in the opposite direction of the main flow. Thus, in the later analysis, a positive wave is directly caused by the movement of components like valves, whereas a negative wave is directly generated by the positive wave arriving at the next node. Defining $E = \frac{H-H_0}{2} - \frac{a}{2g}(v-v_0)$ and $e = \frac{H-H_0}{2} + \frac{a}{2g}(v-v_0)$ as transfer parameters of positive and negative energy, summing and differing E and e give Equations (5) and (6), respectively.

$$\frac{H-H_0}{2} + \frac{a}{2g}(v-v_0) = \left[\frac{H-H_0}{2} + \frac{a}{2g}(v-v_0)\right]_B - \frac{\lambda L_C}{4gd}(v|v| - v_0|v_0|) \tag{3}$$

$$\frac{H-H_0}{2} - \frac{a}{2g}(v-v_0) = \left[\frac{H-H_0}{2} - \frac{a}{2g}(v-v_0)\right]_C - \frac{\lambda L_B}{4gd}(v|v| - v_0|v_0|) \tag{4}$$

$$H - H_0 = E + e \tag{5}$$

$$Q - Q_0 = \pm\frac{g}{aA}(e - E) \tag{6}$$

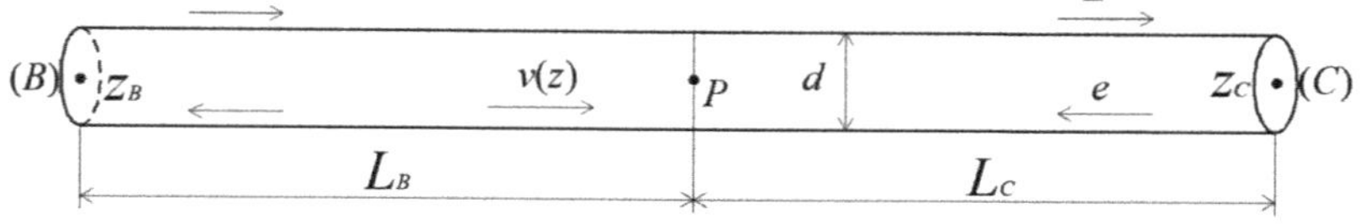

Figure 2. Propagation of pressure wave in pipeline.

In Equation (6), "−" is used when the pressure wave propagation direction is the same as the mainstream and "+" otherwise. Knowing the head, flow velocity, and wave velocity at the initial moment, the flow information at any moment can be solved by Equations (5) and (6). Substituting E and e into Equations (3) and (4) yields Equations (7) and (8), which describe the relationship among energy transfer parameters.

$$e(z-at) = e_B - \frac{\lambda L_B}{4gd}(v|v| - v_0|v_0|) \tag{7}$$

$$E(z+at) = E_C + \frac{\lambda L_C}{4gd}(v|v| - v_0|v_0|) \tag{8}$$

2.2. Pipeline with Varying Diameter and Bifurcated Conduit

Figure 3a shows a pipe with different diameters, where the flow direction is from A_1Q_1 to A_2Q_2. Part of wave energy E_S keeps propagation towards A_2Q_2, while the other part e_S travels in the opposite direction. Consequently, energy transfer parameters at the change in pipe diameter can be written as Equations (9)–(12).

$$E_1 = \frac{H-H_0}{2} - \frac{a}{2g}(v_1 - v_{1,0}) \tag{9}$$

$$E_2 = \frac{H-H_0}{2} - \frac{a}{2g}(v_2 - v_{2,0}) \tag{10}$$

$$e_1 = \frac{H-H_0}{2} + \frac{a}{2g}(v_1 - v_{1,0}) \tag{11}$$

$$e_2 = \frac{H-H_0}{2} + \frac{a}{2g}(v_2 - v_{2,0}) \tag{12}$$

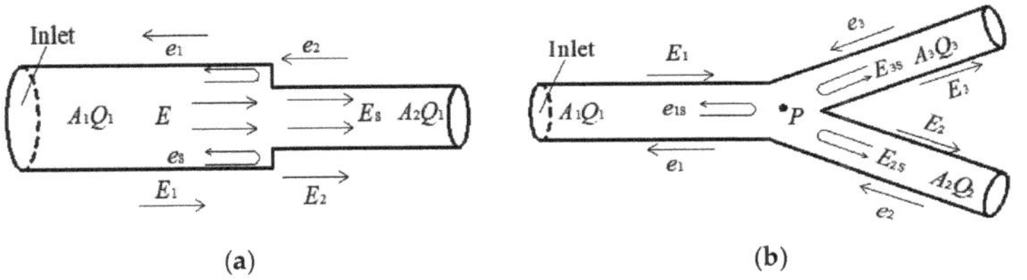

Figure 3. Parameters of transfer energy in pipeline with varying diameter and bifurcated conduit. (**a**) A pipe with different diameters; (**b**) Bifurcated conduit.

E_1 and e_2 come from the previous node and can be solved directly using Equations (7) and (8). However, e_1 and E_2 are caused by the reflected waves. Solving these four equations jointly (Equations (9)–(12)) results in Equations (13) and (14).

$$e_1 = \frac{A_1 - A_2}{A_1 + A_2}E_1 + \frac{2A_2}{A_1 + A_2}e_2 \tag{13}$$

$$E_2 = \frac{2A_1}{A_1 + A_2}E_1 + \frac{A_2 - A_1}{A_1 + A_2}e_2 \tag{14}$$

Figure 3b displays a bifurcated conduit, where the flow direction is from A_2Q_2 to A_1Q_1 and A_3Q_3. Forward waves E_1, E_2, and E_3 move in the same direction as the mainstream, while the reflected waves e_1, e_2, and e_3 move in the opposite direction. e_{1S}, E_{2S}, and E_{3S} are parts of the reflected waves caused by E_1, e_2 and e_3, respectively, when E_1, e_2, and e_3 pass through point P. E_1, e_2, and e_3 come from the previous node, so they can be solved using Equations (7) and (8). e_1 is a confluence of e_2 and e_3, while E_1 is dispersed into E_2 and E_3. Thus, e_1, E_2, and E_3 can be solved using Equations (15)–(17), which are derived from Equations (9)–(12).

$$e_1 = \frac{A_1 - A_2 - A_3}{A_1 + A_2 + A_3}E_1 + \frac{2A_2}{A_1 + A_2 + A_3}e_2 + \frac{2A_3}{A_1 + A_2 + A_3}e_3 \tag{15}$$

$$E_2 = \frac{2A_1}{A_1 + A_2 + A_3}E_1 + \frac{A_2 - A_1 - A_3}{A_1 + A_2 + A_3}e_2 + \frac{2A_3}{A_1 + A_2 + A_3}e_3 \tag{16}$$

$$E_3 = \frac{2A_1}{A_1 + A_2 + A_3}E_1 + \frac{2A_2}{A_1 + A_2 + A_3}e_2 + \frac{A_3 - A_1 - A_2}{A_1 + A_2 + A_3}e_3 \tag{17}$$

2.3. Boundary Conditions of Pump Turbine

The pump turbine is the core of a PSP, and the complete characteristic curves of the unit should be considered when determining boundary conditions. It is worth mentioning that this work also considered the energy transfer parameters from the upstream and downstream. The following Equations (18)–(20) belong to the improved Suter transfer method, used to quantitatively characterize the curves of a unit [38].

$$WH(x,y) = \frac{y^2}{(n_{11}/n_{11r})^2 + (Q_{11}/Q_{11r})^2} = \frac{h}{N^2+q^2}y^2 \tag{18}$$

$$WM(x,y) = \frac{M_{11}+k_1}{M_{11r}}y = \left(\frac{m}{h} + \frac{k_1}{M_{11r}}\right)y \tag{19}$$

$$x = \begin{cases} \arctan\left(\frac{Q_{11}/Q_{11r}+k_2}{n_{11}/n_{11r}}\right) = \arctan\left(\frac{q+k_2\sqrt{h}}{N}\right), & N \geq 0 \\ \pi + \arctan\left(\frac{Q_{11}/Q_{11r}+k_2}{n_{11}/n_{11r}}\right) = \pi + \arctan\left(\frac{q+k_2\sqrt{h}}{N}\right), & N < 0 \end{cases} \tag{20}$$

The positive wave near the upstream side of the pump turbine does not reach the next node at the initial time, so it has no reflected waves, meaning $e_1 = 0$. Similarly, the pump turbine near the downstream has $e_2 = 0$. Combining $e_1 = 0$ and $e_2 = 0$ with Equations (5) and (6) yields Equations (21) and (22). Ignoring the main losses of the pump turbine, such as hydraulic, volume, and mechanical losses, $H_t = H_1 - H_2$. Substituting Equations (21) and (22) into $H_t = H_1 - H_2$ results in Equation (23), which is the expression of the pump turbine boundary.

$$H_1 - H_{1,0} = 2e_1 - \frac{a(Q_1 - Q_{1,0})}{A_1 g} \tag{21}$$

$$H_2 - H_{2,0} = 2e_2 + \frac{a(Q_2 - Q_{2,0})}{A_2 g} \tag{22}$$

$$hH_r = 2(e_1 - e_2) - \frac{a_1}{A_1 g}(qQ_r - Q_{t0}) + \frac{a_2}{A_2 g}(qQ_r - Q_{t0}) + H_{1,0} - H_{2,0} \tag{23}$$

Equation (23) uses subscripts 1 and 2 to distinguish the parameters near the upstream side and downstream side, while the subscript 0 represents the initial time.

3. Validation of WHED in an HP

The WDS of one HP is built in the lab, as shown in Figure 4a. The arrangement consists of one pipe and three machines, and the schematic is shown in Figure 5 and the pipeline parameters are shown in Table 1. The placement of the pressure sensors is described in a previous study [31], and three valves are used to control the flow. To accurately capture the dynamic variation of water hammer pressure in the pressure pipeline, the physical model utilizes both pressure sensors and a high-speed camera to collect pressure and surge water level data in the surge tank. More than 40 pressure sensors are installed in the test rig. This study primarily focuses on the validation of calculations using the pressure sensors H_1, H_2, and H_3, located at the downstream end of pressure pipelines No. 1 to No. 3, as shown in Figure 4a, along with the pressure sensor H_B at the bottom of the surge tank. Additionally, the surge water level H_W recorded by the high-speed camera is also validated. The following four points are included in this experiment:

(1) The overflow weir regulates the water level in the upper reservoir, and the flow rate is determined by the gauging weir in conjunction with the regulating valve.
(2) A high-speed camera records the changes in waves within the surge tank.
(3) The signal acquisition system monitors from 0 to 350 s.

(4) The wave velocity a is measured by quickly cutting off the water flow using a gate to produce water hammer waves. The pressure sensors at both ends of the pipe record the first wave time t_1, which is used to calculate the water hammer wave velocity ($a = l/t_1$).

(a) **(b)**

Figure 4. The information of experiment. (**a**) The layout of test bench; (**b**) Flow control laws.

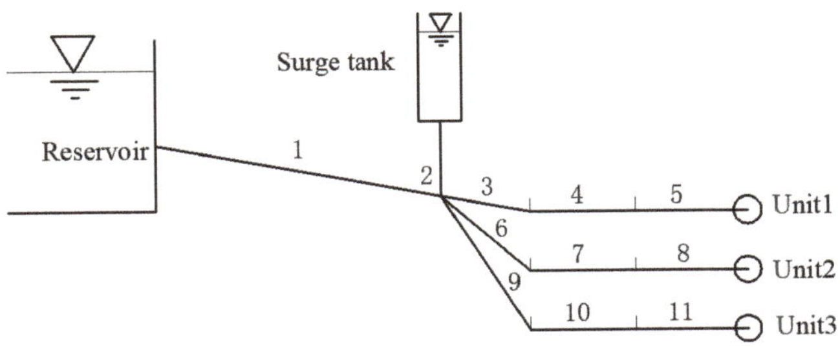

Figure 5. Schematic diagram of HP. (The triangle in the figure represents the water level line, 1–11 represents the pipeline number, and the parameters are shown in Table 1).

Table 1. Pipeline parameters of the WDS.

Number	Length (m)	Pipe Diameter (m)	Area (m^2)	Wave Velocity (m/s)	Roughness
1	493.42	11	94.99	1319	0.015
2	7.8	11	94.99	1319	0.015
3	19.3	6.35	31.65	1157.66	0.012
4	65.55	6.35	31.65	1157.66	0.012
5	53.20	6.35	31.65	1157.66	0.012
6	19.3	6.35	31.65	1157.66	0.012
7	65.55	6.35	31.65	1157.66	0.012
8	53.20	6.35	31.65	1157.66	0.012
9	19.3	6.35	31.65	1157.66	0.012
10	65.55	6.35	31.65	1157.66	0.012
11	53.20	6.35	31.65	1157.66	0.012

In order to ensure that the model experiment can accurately reflect the operation of the actual project, the model test is designed with the Froude criterion. The relationship of the water hammer wave velocity scale (δ_c) and the geometric scale (δ_l) is $\delta_c = \delta_l^{0.5}$. The water hammer wave velocity of the pressure pipe is 1157.66 m/s, which is obtained by considering the hydropower station (1100 m/s). The difference is caused by pipe materials, which is polyvinyl chloride (PVC).

To capture the transient characteristics in the WDS, pressure sensors with a range of 0–100 kPa are arranged at different locations in the pipelines and the surge well, and a video camera is used to record the changing water level in the surge [31]. Pressure signals are collected by a directly attached storage system, and sampling frequency of all sensors is 1000 Hz. The geometric scale of the test model is 42.87.

This paper analyzes the transient characteristics at five special positions, as shown in Figure 4a: H_1, H_2, and H_3 are located at the end of the WDS; H_B is at the bottom of the surge tank; and H_W represents the wave in the surge tank. Detailed information on the four transitions is shown in Table 2, where 163 m and 184 m are the minimum and normal storage levels of the upper reservoir in the HP, respectively.

Table 2. Four transient conditions of HP.

Conditions	Description
ET1	Upstream reservoir—184 m, three units rejecting loads at the same time
ET2	Upstream reservoir—163 m, three units rejecting loads at the same time
ET3	Upstream reservoir—184 m, two units operating at full load, one unit increasing to full load, and then three units rejecting loads at the same time
ET4	Upstream reservoir—163 m, two units operating at full load, one unit increasing to full load, and then three units rejecting loads at the same time

Figure 4b shows the flow control laws under different conditions. The trend of the test results is linear. In Equation (24), R^2 is the linear correlation coefficient, sum of squares of residuals (SSR) and total sum of squares (TSS) are the regression and total sum of squares, respectively. Both SSR and TSS are greater than 0.99, indicating that the accuracy of the test bench control module is high.

$$R^2 = 1 - \frac{SSR}{TSS} \tag{24}$$

The application steps of WHED in HP are shown in Figure 6a, where Ωq, Ωh, and ΩN represent the differences between the predicted and calculated values. Figure 6b illustrates the calculation procedure of the transient conditions detailed in this paper. It should be noted that the red steps in Figure 6b are not included in the research in generating mode (conditions of the turbine model are in the first quadrant).

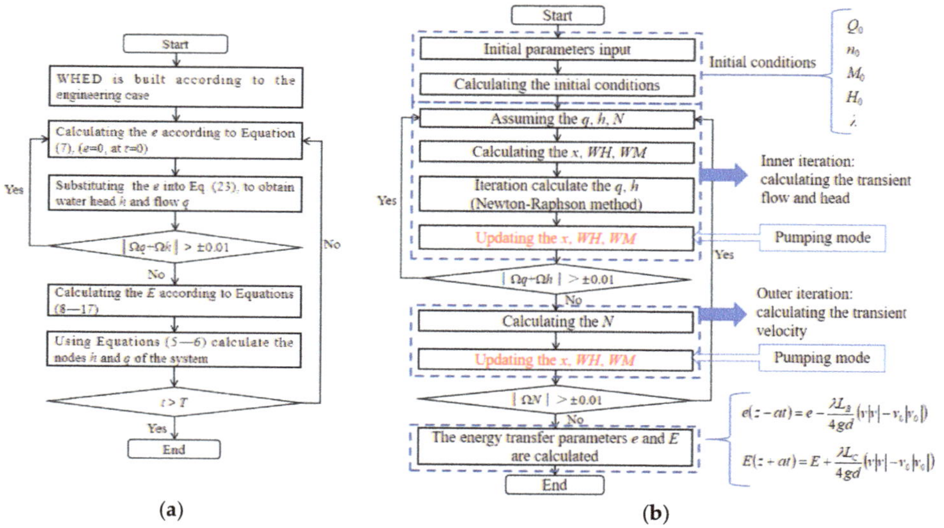

Figure 6. The flow charts of calculation in HP. (**a**) Application steps of WHED; (**b**) The calculation process of transient conditions.

3.1. Comparative Analysis of Numerical and Experimental Results

The calculating data of WHED are consistent with the experiment results. Defining $\Delta_1 = \frac{|WHED-Experiment|}{Experiment} \times 100\%$ to measure the difference between the experiment and WHED, Δ_{1max} of H is only 4.8%. The model pipe in the laboratory is made of PVC for measuring wave velocity, while the hydropower plant uses steel for the WDS. Due to the different elastic moduli of the two materials, their wave velocities differ. In this work, the wave velocity of the steel pipe used in the plant is adopted, which results in the calculated values slightly lagging behind the experimental data. Validation of numerical simulations has been completed in the previous stage [31].

Pressure changes at the end of the WDS and the water level in the surge tank of ET1 are shown in Figure 7a,b. Valve closure causes a rapid rise in pressure at the end of the WDS (H_{max} = 153 m). Part of the water flows into the surge tank, causing H_W to start increasing while H_1, H_2, and H_3 gradually decrease. After four waves, H_B and H_W reach stable values (H_{wmax} = 198 m), with Δ_{1max} of H_w being only 1.1%.

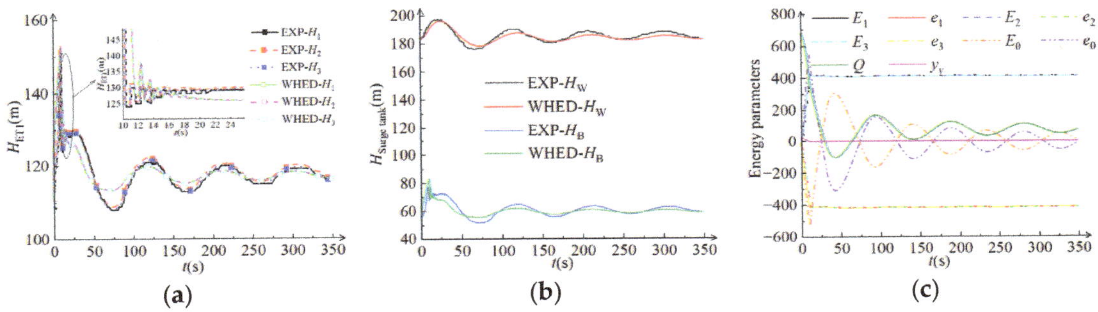

Figure 7. WHED and experiment results at ET1. (**a**) Pressure of H_1, H_2, and H_3; (**b**) Pressure of H_W and H_B; (**c**) Energy transfer parameters.

The changes in valve opening (y_v), flow rate (Q), and energy transfer parameters (E and e) are shown in Figure 7c. The subscripts 1, 2, and 3 indicate the energy transfer parameters at the end of three pipelines of the WDS, while the subscript 0 represents the surge tank. e increases in the negative direction when E reaches the reservoir end. The reflection flow (e) arrives at the end of pipes at t = 1.1, then a positive direct water hammer appears in the pipeline because the valves have not been fully closed, accompanied by a rapid rise in pressure at the ends of pipes. E_0 and e_0 at the surge tank increase in the negative and positive directions, respectively, exhibiting opposite trends compared with E and e at pipe ends, which is consistent with the regulating function of the surge tank. The shrinking fluctuations of E_0 and e_0 after their maximum values are caused by oscillatory changes in flow.

Figure 8 shows that ET2 has a similar situation to ET1, but the maximum H at the end of WDS and H_w are 18 m and 20 m smaller than those in ET1, primarily due to different reservoir levels (H_R). The pressure at each position experiences four waves before stabilization, with Δ_{1max} of H at the end of the WDS being about 4.9% and Δ_{1max} of H_w around 1.8%. The trends of the energy transfer parameters for ET1 and ET2 illustrate the key role played by the surge tank for pressure stabilization within the system.

Valve closure at the pipe ends brings the flow to 0, causing E_1, E_2, and E_3 to first increase and then stabilize. However, the surge fluctuation in the regulating well leads to pulsations in E_0 and e_0, with their pulsation amplitudes gradually shrinking and stabilizing, as shown in Figure 8c. These reasons also apply to the changes in energy parameters in the later figures of this case.

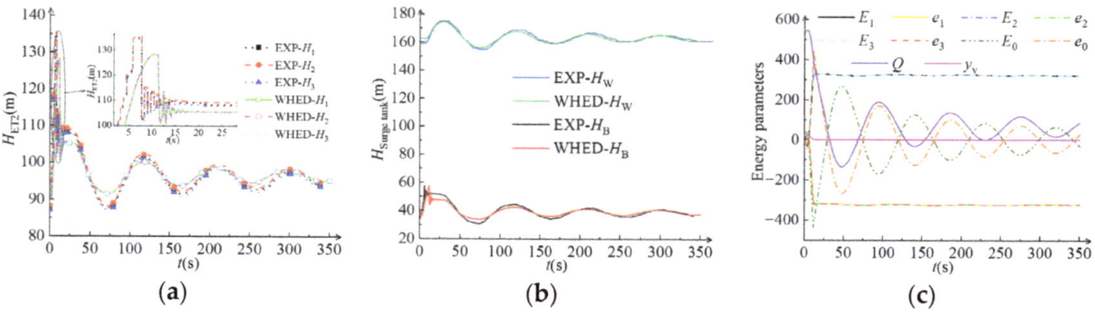

Figure 8. WHED and experiment results at ET2. (**a**) Pressure of H_1, H_2, and H_3; (**b**) Pressure of H_W and H_B; (**c**) Energy transfer parameters.

H at the end of the WDS and H_W and H_B exhibit different change trends in Figure 9. Firstly, the pressure of H_1 decreases due to the larger valve opening from t = 0 to 75 s (H_{1min} = 105 m), accompanied by smaller H_W and H_B. The pressure at H_2 and H_3 experiences small fluctuations from t = 0 to 75 s (H_{min} = 113 m), influenced by the surge tank. e_0 decreases and then increases at the startup of one unit, reflecting the change in water flow in the surge tank from outflow to inflow.

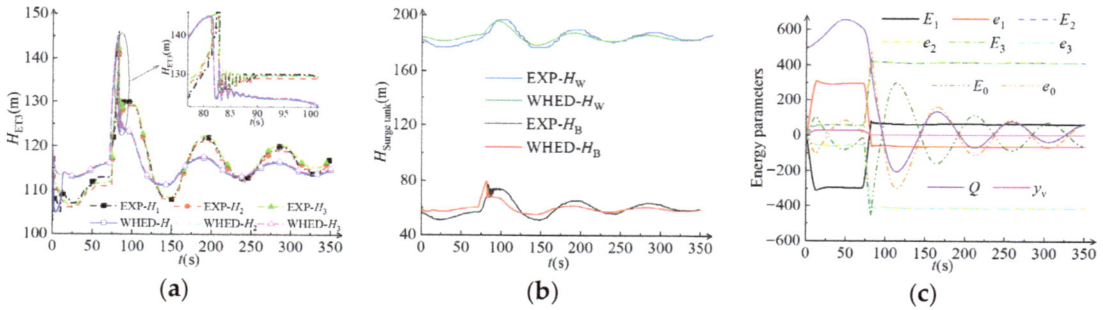

Figure 9. WHED and experiment results at ET3. (**a**) Pressure of H_1, H_2, and H_3; (**b**) Pressure of H_W and H_B; (**c**) Energy transfer parameters.

At t = 75 s, three units start to reject the full load, causing a rapid pressure rise (H_{max} = 146 m) similar to ET1. However, the maximum H is smaller in ET3 than in the simple load rejection condition, as pressure fluctuations at the end of the load increase offset some of the fluctuations at the beginning of load rejection. After t = 75 s, e_0 rises and then falls, with its subsequent three waveforms gradually decaying. The changing trends of WHED and experimental results are essentially the same, with Δ_{1max} of H_1, H_2 (H_3), and H_w being 6.9%, 6.8%, and 3.3%, respectively.

As Figure 10 shows, the changing characteristics of ET4 under both WHED and experimental conditions are similar to ET3. The maximum pressure at the end of the WDS of ET4 is 15 m smaller than in ET3, for the same reason as the difference between ET1 and ET2, caused by lower H_R. Δ_{1max} of H_1, H_2, and H_3 are approximately 4.6%, while Δ_{1max} of H_w is 3.8%. H_w of the surge tank first drops and then rises, with a total of four waveforms appearing in the two stages. Compared with ET3, H_{wmax} is reduced by 19 m.

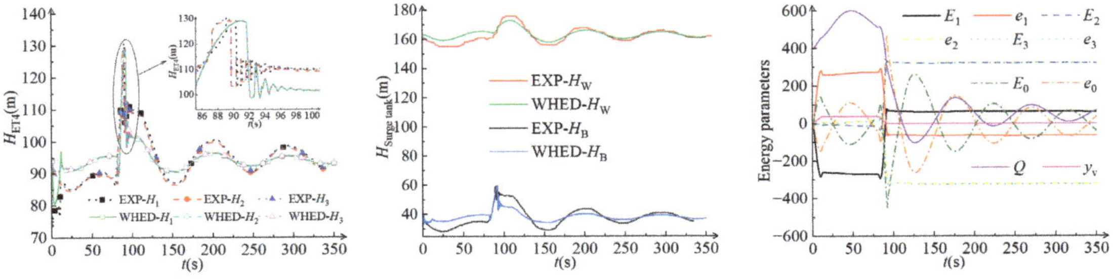

(a) (b) (c)

Figure 10. WHED and experiment results at ET4. (**a**) Pressure of H_1, H_2, and H_3; (**b**) Pressure of H_W and H_B; (**c**) Energy transfer parameters.

WHED is utilized in four transient conditions of HP. Compared to the experimental results, the maximum error Δ_{1max} of each parameter is smaller than 7%, demonstrating the good reliability of WHED. The reservoir level significantly affects the pressure fluctuation amplitude (lower H_R corresponds to smaller maximum values), but it has no effect on the pressure development trends.

3.2. Stability Analysis of Pressure Parameters in WDS of HP

Multidimensional scaling (MDS) is employed to quantitatively analyze the parameters from the transient conditions listed in Table 2, defining u in Equation (25) as a measure of safety, where u_1 is the fluctuating amplitude of parameters as shown in Equation (26), and u_2 denotes the change rate of parameters as expressed in Equation (27).

$$u = c_1 u_1 + c_2 u_2 \tag{25}$$

$$u_1 = \frac{|c_{\max}| - c_r + |c_{\min}| - c_r}{2} \tag{26}$$

$$u_2 = \frac{c_{\max} - c_0}{t} \tag{27}$$

In the above two expressions, $c_{\max}$, $c_{\min}$, and c_r are the maximum, minimum, and rated values of H_1, H_2 (H_3), and H_w, respectively. $c_1 = \frac{\overline{u}_1}{\overline{u}_1+\overline{u}_2}$, $c_2 = \frac{\overline{u}_2}{\overline{u}_1+\overline{u}_2}$, $\overline{u}_1$ and $\overline{u}_2$ are the average values of u_1 and u_2. A larger u_1 indicates a greater change in parameters, while a bigger u_2 signifies a faster change in parameters. Table 3 lists the calculated results of u_1, u_2, and u for the transitions in Table 2.

Table 3. u_1, u_2, and u of four transitions in Table 2.

Conditions	H_1			H_2 (H_3)			H_w		
	u_1	u_2	u	u_1	u_2	u	u_1	u_2	u
ET1	19.51	3.71	17.14	19.61	3.71	16.91	8.82	0.49	8.49
ET2	19.65	3.56	17.24	19.05	3.56	16.42	9.09	0.42	8.74
ET3	19.92	3.88	17.51	15.84	3.83	13.79	7.17	0.12	6.89
ET4	22.06	3.65	19.29	18.92	3.72	16.34	6.84	0.17	6.57

The u_1 of H_1 follows the order ET4 > ET3 > ET2 > ET1, demonstrating that the load regulation mode significantly influences the flow regime. During ET3 and ET4, where one unit increases to full load and then rejects the full load, the u_1 of wave superposition pulsation in these two phases is larger compared to ET1 and ET2. The magnitude pattern of u_2 for H_1, H_2, and H_3 is consistent with u_1. Notably, u_1 and u_2 of H_1 at lower reservoir levels (ET2 and ET4) are higher, while u_1 and u_2 of H_2 and H_3 show the opposite trend. The u values of H_1 are consistently higher than the u values of H_2 and H_3, which indicates that H_1 has stronger fluctuations than H_2 and H_3. The difference in u from H_1, H_2, and H_3 in ET1 and ET2 conditions is less than 1, while the difference in u from H_1, H_2, and H_3 in ET3 and ET4 conditions is about 3, which illustrates that the combined condition (first increasing and then decreasing the load) has a larger pressure fluctuation compared with a single process with load decrement. This further highlights the important impact of load regulation patterns on system safety.

A comprehensive parameter $\overline{u}$ is defined as the weighted average values of q, h, and N in Equation (28). Here, $\overline{u}_{H_1}$, $\overline{u}_{H_2(H_3)}$, and $\overline{u}_{H_w}$ are the average values of u from H_1, H_2 (H_3), and H_w. $c = \overline{u}_{H_1} + \overline{u}_{H_2(H_3)} + \overline{u}_{H_w}$, and the calculating results of these quantities are presented in Table 4. From a comprehensive perspective, the safety order of the conditions in Table 1 is ET3 > ET2 > ET1 > ET4. ET3 exhibits the best performance, while ET4 represents the worst case, with the latter value of $\overline{u}$ being reduced by 1.67 compared to the former condition.

$$\overline{u} = \frac{\overline{u}_{H_1}}{c} + \frac{\overline{u}_{H_2(H_3)}}{c} + \frac{\overline{u}_{H_w}}{c} \tag{28}$$

Table 4. The weighted values of H_1, H_2 (H_3), and H_w.

Conditions	$\overline{u_{H_1}/c}$	$\overline{u_{H_2(H_3)}/c}$	$\overline{u_{Hw}/c}$	$\overline{u}$
ET1	7.34	6.60	1.53	15.47
ET2	7.13	6.40	1.52	15.05
ET3	7.46	5.38	1.26	14.10
ET4	8.22	6.37	1.18	15.77

3.3. Coupled Computation of 1D WHED with 3D Numerical Simulation

The one-dimensional water hammer energy difference calculation method is coupled with the three-dimensional numerical simulation method to calculate the time-dependent variations of flow parameters in the diversion system of a hydropower station. The flow rate at the inlet of the diversion pipe and the pressure at the end of the pressure pipeline are calculated using the one-dimensional energy difference method, which then serves as the boundary conditions for the 3D numerical simulation. A 3D model of the diversion system is established, and meshing is performed. As shown in Figure 11a, five grid configurations (2.6 million, 4.15 million, 5.78 million, 6.0 million, 6.3 million) were tested for grid-independence validation, using the ratio of H/HEXP as the evaluation criterion (where H and HEXP represent the numerical and experimental heads, respectively). When the grid count reached 5.78 million, the H/HEXP ratio stabilized.

A steady state was achieved by filling the diversion system with water, a task completed in previous work [30]. Based on this, the coupled calculation was performed by transferring data from the 1D water hammer energy difference method to the 3D numerical simulation in a one-way manner, while monitoring the surge level in the regulating well during the 1D–3D coupled calculation. Figure 11b presents the comparison between the one-dimensional and three-dimensional coupled results and the model test results for four experimental conditions. The water level trends in the regulating well are largely consistent. The maximum computational deviation for the four experimental conditions is approximately 1.9%, indicating good coupling performance.

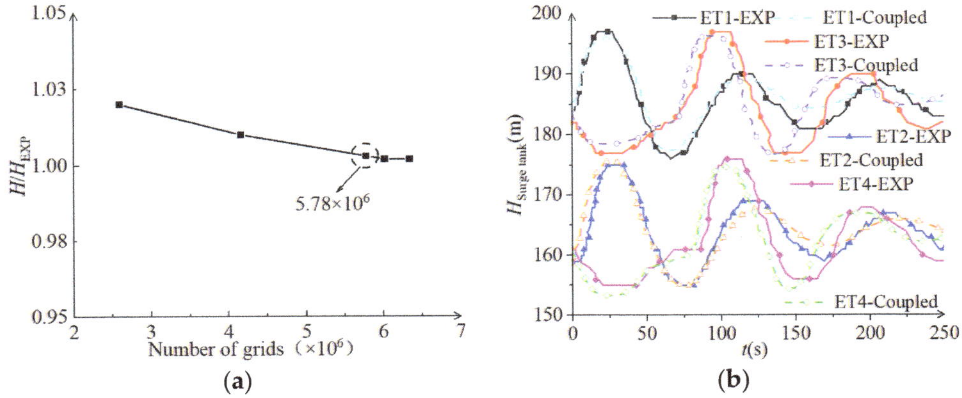

Figure 11. The 1D WHED coupling 3D numerical simulation. (**a**) Grid-independence validation; (**b**) Coupling performance validation.

The turbulence kinetic energy (TKE) in the diversion system for each experimental condition is shown in Figure 12a–d, with each image depicting the moment when the water level in the regulating well reaches its maximum value. Conditions ET1 and ET2 represent load shedding scenarios, with the turbulence intensity in ET2 being more severe than in ET1. Conditions ET3 and ET4 involve first increasing load and then shedding load, where the turbulence intensity in ET4 is more pronounced than in ET3. This indicates that turbulence in the regulating well is more intense under low water level transition conditions. High-TKE regions are concentrated around the impedance holes, where the turbulence intensity is most severe. In all four conditions, irregular flow extends from the impedance hole into the regulating well, with turbulence intensity gradually decreasing as it is influenced by the water pressure in the regulating well. The maximum turbulent kinetic energy in the low-water conditions, ET2 and ET4, is higher than that in the high-water conditions, ET1 and ET3, with the maximum TKE value being approximately twice that of ET1 and ET3.

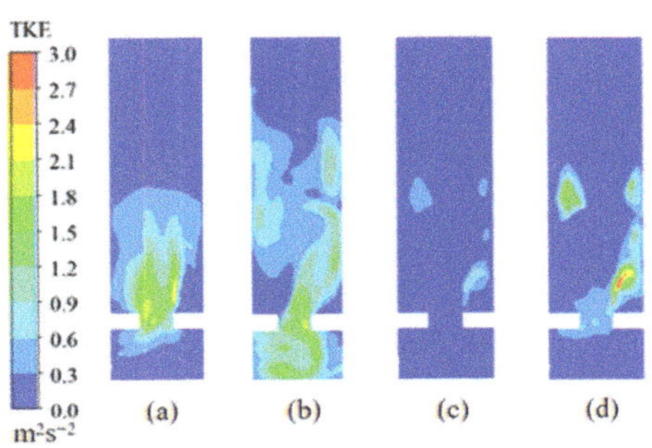

Figure 12. TKE distribution in regulating well under 4 experimental conditions. (**a**) ET1; (**b**) ET2; (**c**) ET3; (**d**) ET4.

4. Validation and Application of WHED in a Pumped Storage Plant

An operating PSP in China is depicted in Figure 13. Since its commissioning and operation until the end of 2020, this PSP has consumed a total of 9.578 billion kW·h of clean energy and has generated a total of 7.833 billion kW·h of electricity. Eight frequently used transient conditions of the PSP case are investigated using WHED, as shown in Table 5. The validation cases include critical working conditions such as load rejection under generating mode (TC0) and accidental power outage under pumping mode (PC0). This Chinese PSP includes upper and lower reservoirs, two water diversion systems with two upstream surge tanks, four units, and one downstream surge tank on the pipeline between the unit and lower reservoir. Both MOC and WHED are employed to calculate the parameter changes under transient conditions. The computational programs for MOC and WHED were completed in a previous study by the authors [39]. The operation process of WHED and the calculation procedure of transitions are shown in Figure 6.

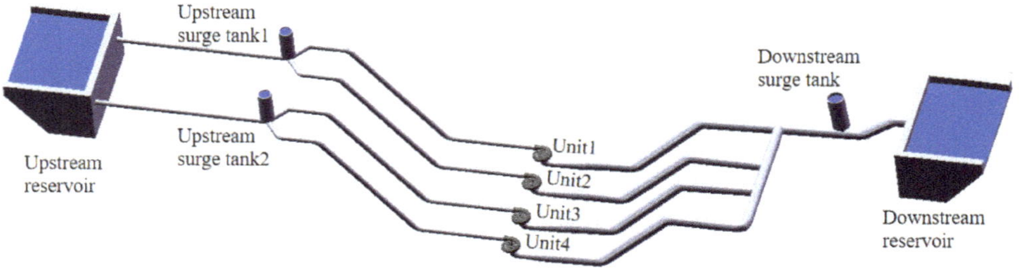

Figure 13. Layout of one PSP in China.

Table 5. Typical transitions.

Designation	Transitions	Description
TC0	Load rejection (Generation mode)	Upstream reservoir—normal water level Initial state—the rated condition Load rejection—guide vanes spend 15 s from 100% to 10% opening
TC1	Load increment (Generation mode)	Upstream reservoir—normal water level Initial state—generating mode with 30% load Load increment—guide vanes spend 20 s from 30% to 100% opening
TC2	Load reduction (Generation mode)	Upstream reservoir—normal water level Initial state—the rated condition Load reduction—guide vanes spend 15 s from 100% to 30% opening
TC3	Shutdown (Generation mode)	Upstream reservoir—normal water level Initial state—the rated condition Shutdown—guide vanes spend 15 s from 100% to 10% opening
TC4	Startup (Generation mode)	Upstream reservoir—normal water level Initial state—shutdown Shutdown—guide vanes spend 20 s from 10% to 100% opening
PC0	Power outage (Pumping mode)	Downstream reservoir—normal water level Initial state—the rated condition Shutdown—guide vanes spend 15 s from 50% to 10% opening
PC1	Shutdown (Pumping mode)	Downstream reservoir—normal water level Initial state—the rated condition Shutdown—guide vanes spend 40 s from 50% to 10% opening
PC2	Startup (Pumping mode)	Downstream reservoir—normal water level Initial state—shutdown Shutdown—guide vanes spend 12 s from 10% to 50% opening

4.1. Validation of Dangerous Working Conditions

Figure 14a,c present the complete characteristic curves of the pump turbine belonging to the PSP shown in Figure 13, while Figure 14b,d display the modified forms using the Suter method, which will be used in subsequent calculations. S_1 in Figure 14a represents the initial point of TC0, corresponding to the normal water level in the upper reservoir and the rated condition of the turbine. During TC0, guide vanes take 15 s to move from y_0 to

10% y_0, following the guide vane closure law $y = 1 - 0.06t$. S_2 in Figure 14a marks the initial state of PC0, where the lower reservoir is at the normal water level and the pump operates under rated conditions. In PC0, the guide vanes take 15 s to move from 50% y_0 to 10% y_0, with the guide vane closure law $y = 0.5 - 0.0267t$. For PC0, a braking mechanism can be used to stop the hydraulic machine quickly in emergency situations, although shutting down by guide vane closure is a comparatively longer process.

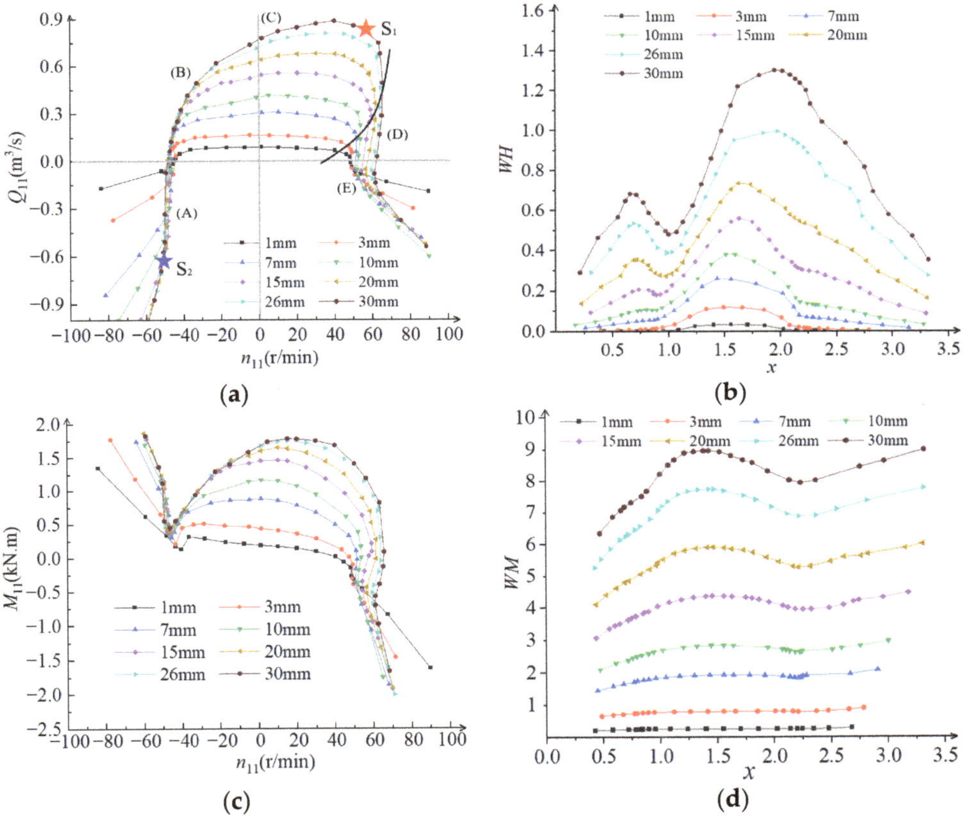

Figure 14. The complete characteristic curves of pump turbine. (**a**) n_{11}-Q_{11}; (**b**) x-WH; (**c**) n_{11}-M_{11}; (**d**) x-WM.

Figure 15a demonstrates the fluctuations in rotation speed, torque, and mass-flow during TC0. The flow continues to decrease until it reaches 0 at $t = 8.9$ s. Subsequently, a reflux appears in the system, reaching a maximum value of $q_{max} = -0.285$, and it takes 30 s to stabilize at $q = 0.155$. The rotation speed increases sharply and then decreases ($N_{max} = 1.47$ at $t = 7$ s) due to the unit experiencing runaway before braking during the load rejection period. The pressure initially rises and then falls ($h_{max} = 1.79$, $h_{min} = 0.66$). The first wave ends at $t = 19$ s, followed by small pressure fluctuations around $h = 1$. These fluctuations persist for about 45 s, as the unit's operating point repeatedly passes through the braking zone and the reverse pump zone at the end of the load rejection process. In addition, the water level pulsation causes the pressure to converge more slowly compared to the flow and rotation speed.

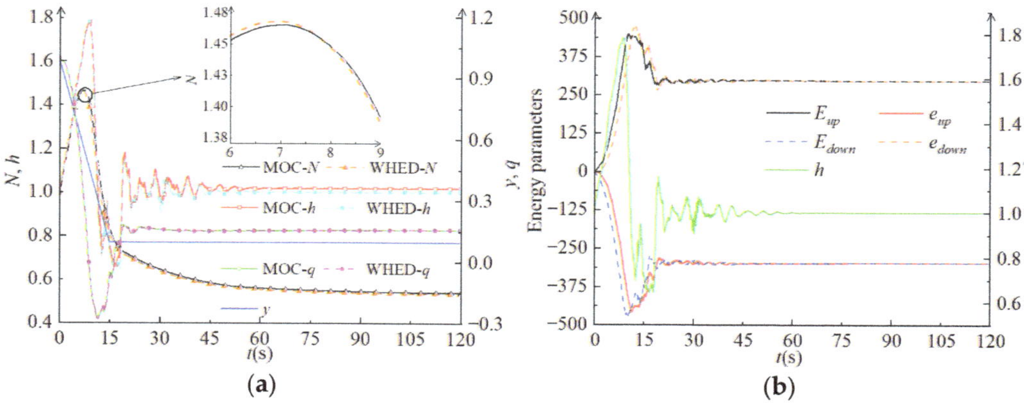

Figure 15. The results of TC0 condition. (**a**) Rotation speed, flow, and pressure; (**b**) Energy transfer parameters.

The vector sum of positive and negative energy transfer parameters corresponds the pressure change between the initial moment and any given time (Equation (5)). Figure 15b illustrates that the forward and reflected waves in the pipelines exhibit similar change patterns, with initial upward trends followed by downward tendencies, ultimately leading to the stabilization of all variables. The differences in parameter values and propagation times are primarily attributed to variations in pipe lengths and the discrepancies in the energy carried by each pressure wave. For instance, as shown in Figure 15b, near the upstream region, $E_{\mathrm{up}} > 0$, $e_{\mathrm{up}} < 0$, and the vector sum of E_{up} and $e_{\mathrm{up}} > 0$, indicating the occurrence of a positive water hammer. E_{up} reaches its maximum value at $t = 10.1$ s, coinciding with the first peak in pressure h. E_{up} decreases immediately after $t = 10.1$ s. The change in e_{up} is similar to E_{up}, but with an opposite value, causing their vector sum to gradually declines, indicating a pressure reduction. e_{up} reaches its maximum at $t = 11.8$ s, corresponding to the first trough in pressure h. After $t = 12.4$ s, minor fluctuations appear in the curves of E_{up} and e_{up}, which correspond to small fluctuations in h. The vector sum of E_{up} and e_{up} equals 0 at $t = 60$ s, demonstrating that the system has reached a new stable operating state. The water hammer phenomenon on the downstream side is completely opposite to that on the upstream side. On the upstream side, the initial water hammer pressure wave moves counter to the mainstream direction, whereas on the downstream side, it aligns with the mainstream direction.

Figure 16 illustrates the continuous decrease in rotation speed under the PC0 condition. After $t = 43.3$ s, the unit enters the reverse pump zone, where it begins rotating in the opposite direction, with its rotation speed increasing to around $N = 0.557$. The initial sharp drop in rotational speed is caused by the disconnection between the motor and the unit. Subsequently, the unit begins to rotate in the opposite direction, leading to increased reflux, which causes the unit to enter the turbine zone in reverse and accelerate toward a runaway condition. The flow declines from the beginning until $t = 7.11$ s, when backflow appears at the unit outlet. The main cause of this backflow at $t = 7.11$ s is the unit entering the braking zone, where the rotation speed reaches 0. The maximum backflow is $q_{\max} = 0.3$ at $t = 10.4$ s and eventually stabilizes at approximately $q = 0.16$.

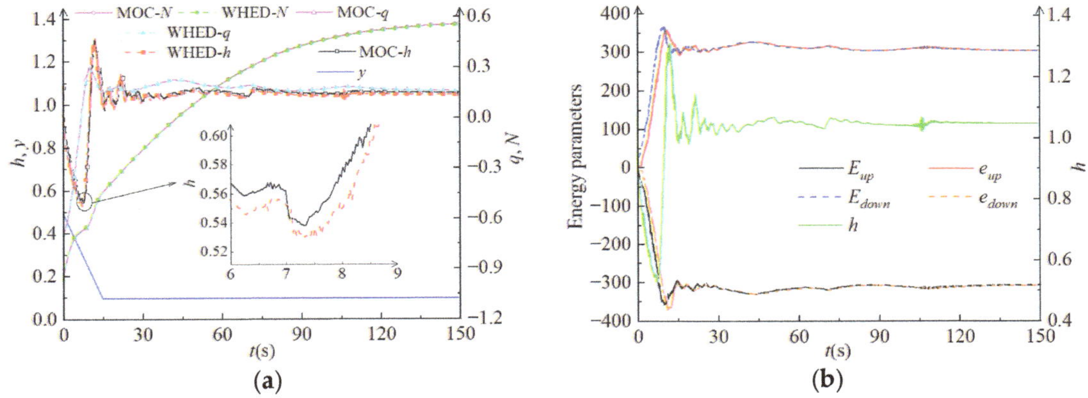

Figure 16. The results of PC0 condition. (**a**) Rotation speed, flow, and pressure; (**b**) Energy transfer parameters.

Figure 16b shows that $E_{up} < 0$ and $e_{up} > 0$, with the vector sum of E_{up} and $e_{up} < 0$, indicating the presence of a negative water hammer in the system. E_{up} decreases to a minimum at $t = 9.93$ s, while e_{up} initially increases, causing their sum to rise slowly and resulting in an upward trend in pressure. At $t = 11.28$ s, corresponding to the maximum value of e_{up}, the pressure h reaches its peak value. Small fluctuations in E_{up} and e_{up} are observed under both critical conditions (TC0: −294.4 to −327.4 and PC0: 294.5 to 328.2). Following the first waves of E_{up} and e_{up}, the pressure fluctuates within the range of 0.985 to 1.145 during this stage. The sum of E_{up} and e_{up} reaches 0 after 106.5 s under the PC0 condition, indicating that the system has reached a new stabilized state.

To measure the differences between MOC and WHED, the relative error $\Delta_2 = \frac{|WHED-TLM|}{TLM} \times 100\%$ is calculated as the absolute difference between MOC result and WHED result, divided by MOC. Two partial enlargements in Figures 15a and 16a show N ($\Delta_2 = 0.2\%$) at TC0 and h ($\Delta_2 = 1.8\%$) at PC0 calculated by WHED and MOC, respectively. Table 6 shows the other parameters' Δ_2. The minor differences at TC0 are: $\Delta_{2N} = 0.7\%$, $\Delta_{2h} = 1.9\%$, and $\Delta_{2q} = 2.8\%$. The minor differences at PC0 are: $\Delta_{2N} = 0.5\%$, $\Delta_{2h} = 1.8\%$, and $\Delta_{2q} = 1.1\%$. Table 2 also elaborates that the Δ_2 values of N_{min}, h_{min}, and q_{min} are larger at TC0 than that at PC0, whereas the Δ_2 values of N_{max}, h_{max}, and q_{max} are vice versa. The Δ_2 is smaller than 3%, which interprets the credibility of WHED. Furthermore, WHED has several advantages over MOC:

(1) WHED adopts energy transfer parameters from the initial and calculating time, using Equations (3)–(8) to derive boundary equations that solve the system. However, MOC uses two adjacent nodes at the previous moment to build the characteristic functions for the calculating moment. The functions constructed by each pair of nodes are generally different, making MOC more complicated than WHED.

(2) WHED has better timeliness because it only needs to calculate the boundary conditions of the two endpoints of the target segment. By dynamically considering the wave propagation time, WHED allows for a larger time step, significantly improving computational efficiency. Conversely, MOC requires dividing the pipeline into multiple sections and constructing equations with adjacent nodes, requiring a smaller time step and the processing of a large amount of unnecessary nodes.

(3) WHED provides a clear physical interpretation by analyzing changes in energy transfer parameters to explain pressure changes under transient conditions. However,

MOC relies on characteristic equations based on the finite difference method, which do not directly characterize the causes of changes in the system's flow regimes.

Table 6. Comparison between WHED and MOC under TC0 and PC0.

		N_{max}	N_{min}	h_{max}	h_{min}	q_{max}	q_{min}
TC0	WHED	1.469	0.541	1.796	0.673	1.010	−0.277
	MOC	1.465	0.537	1.794	0.659	1.000	−0.285
	Δ_2	0.2%	0.7%	0.1%	1.9%	1.0%	2.8%
PC0	WHED	0.560	−0.957	1.305	0.543	0.301	−0.709
	MOC	0.557	−0.956	1.297	0.533	0.298	−0.717
	Δ_2	0.5%	0.1%	0.6%	1.8%	1.0%	1.1%

4.2. Calculation and Analysis of Generation Conditions

Figure 17a shows the continuous movement of the guide vane ($y = 0.035t + 0.3$) at TC1. Taking the upstream side as an example, the sum of E_{up} and $e_{up} < 0$ indicates the occurrence of a negative water hammer on the upstream side. $E_{up} < 0$ and the water hammer wave is a decompression wave, and the negative waves are reflected from positive waves ($e_{up} > 0$ represents boost waves). The head rises and falls through four waves, with the water hammer wave causing changes in each parameter. The pressure minimized at t = 8.6 s (H_{upmin} = 355.6 m), and positive waves always appear before negative waves. The sum of E_{up} and e_{up} begins to increase gradually, which signifies pressure H_{up} growth. Finally, the sum of E_{up} and e_{up} reaches 0, indicating that the system is in a stable state. The propagations of pressure waves on both the upstream and downstream sides are similar. However, the changing trends of energy parameters and pressure near the downstream side are opposite to those near the upstream side, of which E_{max}, e_{max}, H_{est}, and the times to peaks or valleys are shown in Table 7. TC1 (70% load increase) and TC0 (90% load rejection) have opposite changing rules of energy transfer parameters, wherein E and e of TC1 reach stable states after the maximums, but E and e of TC0 keep fluctuating due to the machinery jumping between turbine and pump working areas. Thus, TC0 has more pressure waves than TC1, and the minimum of TC0 is 11.8% smaller than that of TC1. For instance, the maximum values of E_{up} and e_{up} at TC1 are about 0.59 times those belonging to TC0, and the minimum pressure is about 1.32 times higher.

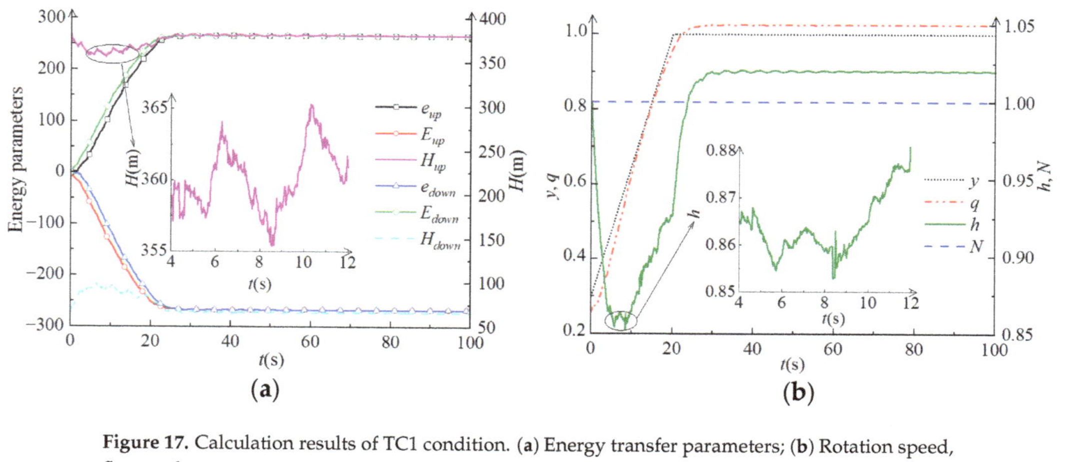

Figure 17. Calculation results of TC1 condition. (**a**) Energy transfer parameters; (**b**) Rotation speed, flow, and pressure.

Table 7. Energy transfer parameters near the upstream and downstream sides at TC1.

TC1	E_{max}	e_{max}	H_{est} (m)	T (s)
Upstream side	−265.8	265.7	355.6	8.6
Downstream side	265.7	−265.2	99.1	9.3

As shown in Figure 17b, the flow rate q increases with the guide vane opening, reaching a maximum at t = 24 s (q_{max} = 1.02). The unit's rotation speed N remains constant, similar to TC2. Two small fluctuations in h and H_{up} occur within the range of 4–12 s. The pressure changing trends on the upstream and downstream sides are opposite, as h can be calculated by Equation (23). The pressure reaches its minimum at t = 8.2 s (h_{min} = 0.852), after which it rises to h = 1.02 and basically remains stable. During the load rejection of TC0, the unit transitions through three different conditions: the turbine zone, braking zone, and reverse pump zone. In contrast, TC1 remains within the turbine zone. Consequently, the fluctuation amplitudes of the parameters are smaller at TC1 compared to TC0.

The guide vane closure law at TC2 is $y = 1 - 0.0467t$, as shown in Figure 18a. On the upstream side of the unit, the vector sum of E_{up} and $e_{up} > 0$, indicating the presence of a positive water hammer. $E_{up} > 0$ represents a positive pressure wave, while $e_{up} < 0$ indicates a decreasing wave due to the reflection of E_{up}. H first rises and then falls, reaching a maximum value of H_{upmax} = 407.08 m. Both the sum of E_{up} and e_{up} and the pressure reach their maximum values at t = 15.1 s. Moreover, E_{up} and e_{up} have their extreme values at t = 21.8 s and t = 23.6 s, respectively. Eventually, the sum of E_{up} and e_{up} returns to 0, indicating that the system has entered a new stabilized operating state. The pressure peaks and troughs of the energy parameters near the upstream and downstream sides at TC2 are shown in Table 8. The numeric symbols of E_{max} and e_{max} are influenced by positive and negative waves, which correspond to the increasing directions of energy transfer parameters. TC2 (70% load decrease) and TC0 (90% load rejection) have similar flow changing trends, as well as the developing tendencies of energy transfer parameters. TC2 and TC1 are identical in that the energy transfer parameter mainly undergoes a fluctuation, because both of them are in the turbine area, and they have similar maximums of E and e (about 250).

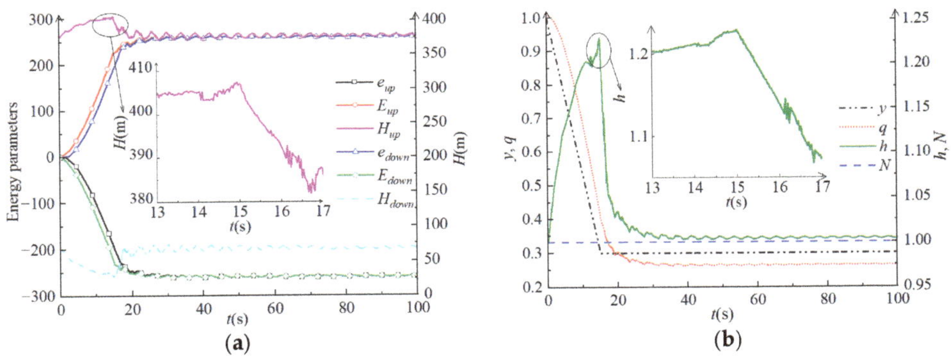

Figure 18. Calculation results of TC2 condition. (**a**) Energy transfer parameters; (**b**) Rotation speed, flow, and pressure.

Table 8. Energy transfer parameters near the upstream and downstream sides at TC2.

TC2	E_{max}	e_{max}	H_{est} (m)	T (s)
Upstream side	258.27	−258.69	407.08	15.1
Downstream side	−257.46	257.63	28.06	15.2

As shown in Figure 18b, the flow rate q decreases with the closing of the guide vane, eventually being reduced to q_{min} = 0.26. The pressure experiences an upward wave (h_{max} = 1.23), ultimately stabilizing about h = 1.0. Both TC1 and TC2 remain in power generation mode, allowing the flow and pressure to converge quickly. TC2 is in the turbine zone like TC1, with the maximum values of E_{up} and e_{up} at TC2 being about 0.57 times those at TC0, and the maximum pressure is reduced by 31.3% compared with TC0.

As shown in Figure 19a, the guide vane remains stationary until t = 11 s. After this point, during TC3, the guide vane closure follows the law $y = 1 - 0.06t$. On the upstream side of the unit, the sum of E_{up} and $e_{up} > 0$, indicating a positive water hammer. $E_{up} > 0$ means the water hammer is a boost wave, with the maximum pressure H_{upmax} = 438.73 m appearing at t = 25.24 s. A buck wave $e_{up} < 0$ is caused by the reflection of E_{up}. Similar to TC2, E_{up} peaks (t = 28.1 s) earlier than e_{up}, leading to a gradual decrease in the vector sum of E_{up} and e_{up} and the pressure. TC3 is a normal shutdown, wherein the unit leaves the power grid after reducing 70% of the load, with the second stage of the remaining 20% load reduction being the same as TC0. Fluctuations of E_{up} and e_{up} lasted for 42 s, with their sum equaling 0. The energy transfer parameters at TC3 are shown in Table 9, with H_{est} appearing 10 s later compared to the TC2 condition, because the guide vanes start to close after a 10 s delay. Furthermore, TC3 exhibits a more pronounced water hammer phenomenon, shedding an additional 20% load compared to TC2, with the E_{max} and e_{max} being about 1.5 times greater than those in TC2.

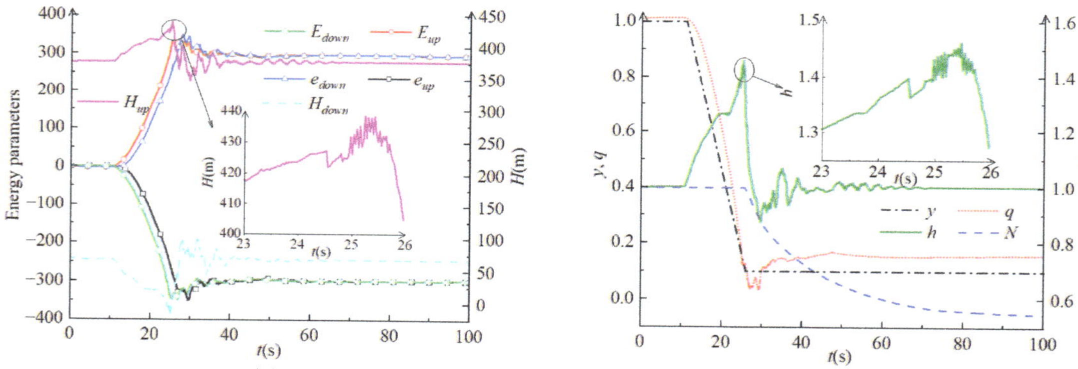

Figure 19. Calculation results of TC3 condition. (**a**) Energy transfer parameters; (**b**) Rotation speed, flow, and pressure.

Table 9. Energy transfer parameters near the upstream and downstream sides at TC3.

TC3	E_{max}	e_{max}	H_{est} (m)	T (s)
Upstream side	347.86	−348.22	438.73	25.24
Downstream side	−349.21	350.68	−17.46	25.47

The parameters changes of the pump turbine are shown in Figure 19b. The flow experiences downward waves (q_{min} = 0.04 at t = 29.5 s), then it gradually increases to a stable value of q = 0.155 over 29 s. The rotation speed remains essentially constant until the guide vane opening reaches 10% y_0. After this point, the unit disconnects from the power grid and enters the braking condition, where the rotation speed decreases. During the initial shutdown period, the pressure experiences a significant increasing wave (h_{max} = 1.45, h_{min} = 0.88), followed by some small fluctuations (0.88–1.07) after t = 27.9 s. Comparing the interval between the maximum and minimum values at TC0 (h_{max} = 1.79, h_{min} = 0.66), the range at TC3 is narrower, which shows that closing the guide vane before disconnecting from the power grid effectively reduces pressure fluctuation amplitudes.

It should be noted that after the peaks and troughs of the energy transfer parameters, their developments are consistent with those at TC0, as the unit has entered the braking zone. Additionally, since only 20% of the load is reduced at TC3, the maximum values of E_{up} and e_{up} are approximately 0.76 times those at TC0.

Figure 20a depicts the two stages of TC4. On the upstream side of unit, $E_{up} < 0$, $e_{up} > 0$, and the sum of E_{up} and $e_{up} < 0$, indicating the presence of a negative water hammer in this side. During the first 30 s, both E_{up} and e_{up} exhibit upward trends, with their sum oscillating between positive and negative, causing the pressure to fluctuate between 353 m and 396 m. In the second stage, as the guide vane rapidly opens, E_{up} and e_{up} experience significant increases, and their sum < 0 amplifies the pressure decrease. E_{up} and e_{up} start to magnify after $t = 35.9$ s, as does the pressure. Ultimately, the absolute values of E_{up} and e_{up} equalize. The energy transfer parameters on the downstream side at TC4 show the opposite circumstance compared to the upstream side, as listed in Table 10. TC4 (startup at generating mode) and TC1 (70% load increase) are load increasing processes, and they have similar trends in E and e. The first 30% load increment at TC4 is the beginning of TC1, hence the energy transfer parameters at TC1 are 21.9% smaller than that of TC4. The last 70% load change at TC4 is same as TC1, and they become stable after a wave.

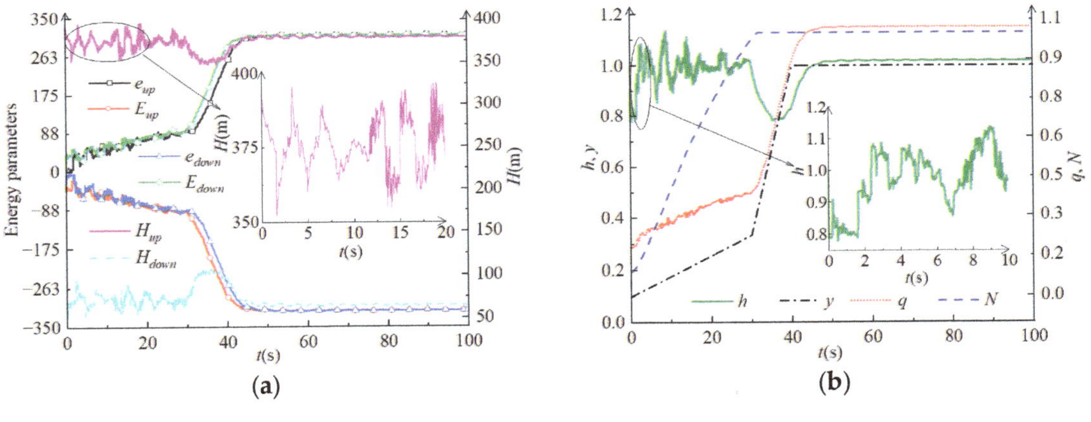

Figure 20. Calculation results of TC4 condition. (**a**) Energy transfer parameters; (**b**) Rotation speed, flow, and pressure.

Table 10. Energy transfer parameters near the upstream and downstream sides at TC4.

TC4	E_{max}	e_{max}	H_{est} (m)	T (s)
Upstream side	−308.49	309.35	348.6	35.87
Downstream side	307.37	−307.57	36.81	36.09

Figure 20b further shows the two stages of TC4. In the first stage ($y = 0.008t + 0.1$), the flow gradually increases ($t = 30$ s, $q = 0.39$), and the rotation speed rises rapidly, stabilizing around $N = 1$. h and H_{up} follow similar trends, experiencing downward waves ($h_{min} = 0.78$ at $t = 0.6$ s). The unit moves from the braking zone to runaway zone, then enters the turbine condition, causing the pressure to undergo five waves ($h_{max} = 1.14$). In the second stage ($y = 0.067t + 0.33$), the guide vane takes 10 s to open from 30% y_0 to 100% y_0, while the rotation speed remains constant because the unit is connected to the power grid in the first stage. However, the flow continues to increase, reaching a stable value of approximately

$q = 1.02$, and the pressure undergoes a downward wave, stabilizing around $h = 1.02$. The second stage is similar to TC1, with the minimum pressure at TC4 being 0.93 times that at TC1. Both TC1 and TC4 involve load increasing conditions, but TC4 experiences a condition switch from the braking zone to the turbine zone in its first stage, hence its final values of E_{up} and e_{up} in the second stage being about 1.15 times greater than TC1.

4.3. Calculation and Analysis of Pumping Conditions

As shown in Figure 21a, the energy transfer parameters on the upstream side of the unit before $t = 20$ s exhibit small fluctuations (−4.03 to −13.46 for E_{up} and 0.75 to 13.04 for e_{up}). At $t = 30$ s, the vector sum of E_{up} and $e_{up} < 0$ represents a negative water hammer. $E_{up} < 0$ signifies a buck wave, and the boost wave ($e_{up} > 0$) appears after reflection of E_{up}. At $t = 49.9$ s, E_{up} peaks while e_{up} is still rising, resulting in a gradual increase in their sum and the pressure. e_{up} reaches its maximum value at $t = 50.8$ s, while the pressure also reaches the maximum after 1.44 s. PC1 and PC0 share some similarities, such as the evolution trends of E_{up} and e_{up}, which exhibit small fluctuations after their peaks (−306.91 to −337.32 for E_{up} and 307.96 to 336.88 for e_{up}). The pressure fluctuation range is 341.7 m to 439.8 m. Finally, the vector sum of E_{up} and e_{up} equals 0. The energy transfer parameters on the downstream side at PC1 show the opposite circumstance compared with that of the upstream side, as listed in Table 11. Comparing the shutdown periods of PC1 and TC3, E_{max} and e_{max} have opposite signs.

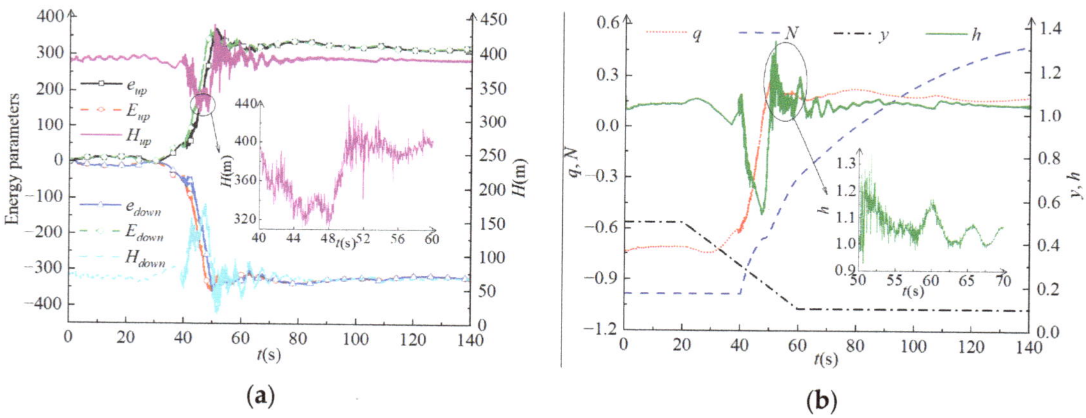

Figure 21. Calculation results of PC1 condition. (**a**) Energy transfer parameters; (**b**) Rotation speed, flow, and pressure.

Table 11. Energy transfer parameters near the upstream and downstream sides at PC1.

PC1	E_{max}	e_{max}	H_{est} (m)	T (s)
Upstream side	−370.59	370.05	311.73	44.97
Downstream side	369.64	−370.45	159.29	42.63

Figure 21 shows two stages at PC1. In the first stage, without guide vane movement before $t = 20$ s, parameters like flow remain essentially unchanged. The guide vane closure law is $y = 0.5 - 0.01t$ in the second stage after $t = 20$ s, leading to a decrease in flow as the guide vane closes, while the rotation speed remains unchanged. The guide vane opening is 30% y_0 at $t = 40$ s, and the power supply is cut off at this moment. Reflux appears at $t = 46.8$ s, reaching a maximum of $q_{max} = 0.28$. The main flow takes about 52 s to stabilize at $q = 0.17$. The rotation speed drops to 0 at $t = 79.4$, after which the reverse flow causes the unit to start rotating in the opposite direction, with a final reverse speed of around

$N = 0.48$. The pressure experiences a downward wave ($h_{min} = 0.55$, $h_{max} = 1.27$), taking 80 s to stabilize at $h = 1.04$. PC0 and PC1 are shutdown transitions under pumping mode, differing in guide vane closure time and power outage time. These differences are reflected in results such as the extreme values of pressure (with h_{min} and h_{max} at PC1 being 1.06 and 0.94 times those at PC0) and the extreme values of E_{up} and e_{up} around ±365 and ±355 at PC0. From the viewpoint of time, the pressure and energy transfer parameters reach their maximum values at PC0 about 40 s earlier than at PC1, due to faster guide vane closure at PC0.

As Figure 22a shows, on the upstream side of the unit, the vector sum of E_{up} and $e_{up} > 0$ represents a positive water hammer. $E_{up} > 0$, and the H_{up} rises to 143% of its original value. $e_{up} < 0$ appears after the reflection from E_{up}, continuing to rise even though E_{up} peaks at $t = 18.1$ s. Their vector sum decreases along with lower pressure. Lastly, the sum of E_{up} and e_{up} returns to 0, indicating the system has reached a new stable state. Table 12 provides information on energy transfer parameters near the downstream side at PC2.

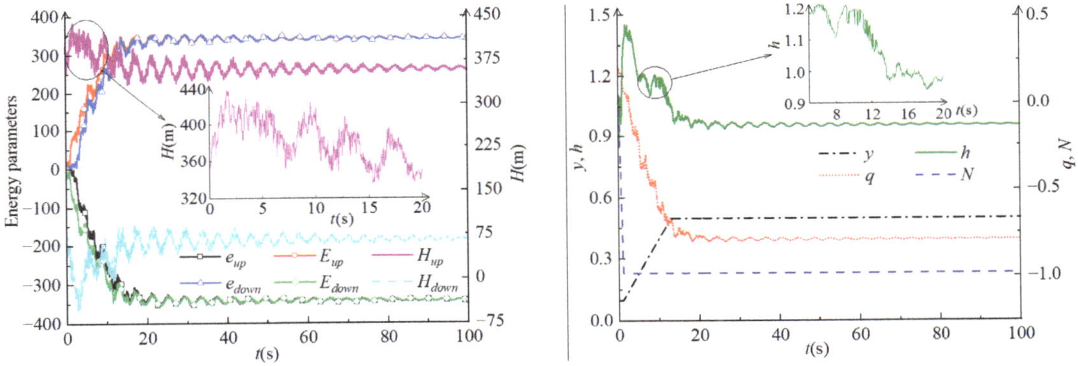

(a) **(b)**

Figure 22. Calculation results of PC2 condition. (**a**) Energy transfer parameters; (**b**) Rotation speed, flow, and pressure.

Table 12. Energy transfer parameters near the upstream and downstream sides at PC2.

PC2	E_{max}	e_{max}	H_{est} (m)	T (s)
Upstream side	348.77	−348.17	438.58	1.83
Downstream side	−344.34	344.12	−50.14	2.58

Figure 22b shows a sharp rise in rotation speed due to motor connection at PC2, with the rotation speed reaching $N = 1$ within 1 s. The pressure quickly reaches its maximum ($h_{max} = 1.43$ at $t = 1.8$ s) after the unit is powered on. h and H_{up} show similar trends, with a small fluctuation at $t = 8$ s. h declines as the guide vanes are further opened, stabilizing at $h = 0.96$ after 55 s. The maximum flow is $q_{max} = -0.8$ at $t = 23.4$ s.

4.4. Stability Analysis of Parameters (Pressure, Flow Rate, and Rotating Speed)

MDS is used to quantitatively analyze the parameters from transient conditions in Table 5. u, u_1, and u_2 are defined by Equations (25)–(27). In these expressions, c_{max}, c_{min}, and c_r are the maximum, minimum, and rated values of q, h, and N, respectively. Table 13 lists the calculated results of u_1, u_2, and u under the transitions in Table 4.

Table 13. u_1, u_2, and u of PSP conditions.

Conditions	q			h			N		
	u_1	u_2	u	u_1	u_2	u	u_1	u_2	u
TC0	0.64	0.11	0.587	0.57	0.09	0.498	0.47	0.07	0.418
TC1	0.39	0.03	0.345	0.08	0.02	0.071	0.00	0.00	0.000
TC2	0.36	0.03	0.327	0.11	0.02	0.097	0.00	0.00	0.000
TC3	0.39	0.06	0.357	0.30	0.03	0.260	0.23	0.01	0.201
TC4	0.38	0.01	0.343	0.17	0.07	0.155	0.42	0.03	0.369
PC0	0.50	0.10	0.460	0.39	0.06	0.341	0.29	0.01	0.254
PC1	0.49	0.04	0.445	0.37	0.03	0.319	0.24	0.01	0.210
PC2	0.49	0.04	0.445	0.24	0.04	0.210	0.50	0.20	0.461

The u_1 of q follows the order TC0 > PC0 > PC1 = PC2 > TC3 = TC1 > TC4 > TC2. Comparing the startup and shutdown conditions of two modes, u_1 of PC1 and PC2 are approximately 25% larger than that of TC3 and TC4, mainly due to more obvious reflux in pumping mode. h, N, and q have similar tendencies in these four conditions, with the pumping mode exhibiting a wider fluctuation range of parameters. The u_2 of h at TC3 and TC4 is larger than that of PC1 and PC2, whereas the u_2 of q and N differ from that of h. The u_2 of q and N are larger at PC2 than at TC4, but they are greater than or equal to PC1 in TC3.

Under generation conditions in Table 5, the parameters of TC0 have the highest u_1, with the maximum and minimum u_1 of q and h differing by factors of about 2 and 7 times, respectively. The maximum and minimum u_1 of N are 0.47 and 0. As for pumping states in Table 5, the u_1 of q and h at PC0 obtain maximum values, whereas the u_1 of N reaches the maximum at PC2. The change pattern of u_2 is similar to u_1, illustrating the consistency of the two prediction results of system safety. u of N in two startup conditions (PC2 and TC4) are the largest, followed by u of q and h. However, u of q in other conditions is larger than u of h, with the smallest being u of N. A comprehensive parameter $\overline{u}$ is defined as the weighted average of q, h, and N in Equation (29).

$$\overline{u} = \frac{\overline{u}_q}{c} + \frac{\overline{u}_h}{c} + \frac{\overline{u}_N}{c} \tag{29}$$

$\overline{u}_q$, $\overline{u}_h$, and $\overline{u}_N$ are the average values of u from q, h, and N, respectively. $c = \overline{u}_q + \overline{u}_h + \overline{u}_N$, and the calculated results of these parameters are presented in Table 14. The order of condition safety is TC2 = TC1 > TC3 > TC4 > PC1 > PC0 > PC2 > TC0. TC1 and TC2 exhibit the best safety, whereas TC0 shows the worst case, with the $\overline{u}$ of TC0 being approximately 2.5 times that of TC1 and TC2.

Table 14. The weighted values of q, h, and N.

Conditions	$\overline{u}_q/c$	$\overline{u}_h/c$	$\overline{u}_N/c$	$\overline{u}$
TC0	0.27	0.13	0.11	0.51
TC1	0.16	0.02	0.00	0.18
TC2	0.15	0.03	0.00	0.18
TC3	0.16	0.07	0.05	0.28
TC4	0.16	0.04	0.10	0.30
PC0	0.21	0.09	0.07	0.37
PC1	0.20	0.09	0.06	0.35
PC2	0.20	0.06	0.12	0.38

5. Conclusions

This work proposes a method called WHED used for guarantee calculation of regulation for HPs and PSPs. The WHED calculation results for HPs show good agreement with experimental data, demonstrating WHED's reliability. Furthermore, WHED and MOC are used to analyze the generating and pumping modes of PSPs, with results revealing WHED's superiority over MOC. Energy parameters are defined to characterize the operational stability of transitions, while MSD is applied to explore the stability of transient parameters.

The key findings of this study include the following three aspects.

(1) A key contribution of this paper is the proposal of WHED: WHED employs energy transfer parameters to characterize system stability, which makes it possible to explain the operations of HPs and PSPs from a physical point of view, and it further breaks the limitation of MOC being limited to mathematical analyses. Thus WHED does not need to take into account the Courant condition in its calculation. Also, WHED can be used with a large time step condition so that it can directly calculate the transient parameters of the target node, improving calculation speed by about 4 times compared with MOC.

(2) The behavior of core parameters in WHED: Negative energy waves are always reflected from positive energy waves, meaning that the negative energy transfer parameter appears later than the positive energy transfer parameter. The energy transfer parameter reflects the energy variation in the system, and the pattern of energy transfer on the upstream side is opposite to that on the downstream side. The regulating well can effectively reduce water hammer pressure, so the energy transfer parameter in the regulating well changes in the opposite direction to that in the pipeline. Larger load adjustments correspond to greater changes in energy transfer parameters and increased system instability. The positive and negative values of the energy transfer parameters indicate different types of water hammer, with positive values corresponding to positive water hammer as pressure increases. The system is stable when the sum of the positive and negative energy transfer parameters equals zero.

(3) This paper verifies the accuracy of WHED by comparing it with model tests and MOC. Due to limitations of the test bench, the wave speed used in the model experiment differs from that in the calculation, resulting in a larger discrepancy in the model test compared to the calculation verification. The smallest error occurs in the TC0 condition, demonstrating that the water hammer energy difference calculation method is highly reliable based on the comparison of results. Additionally, the one-dimensional and three-dimensional coupled calculations highlight the broad applicability of WHED in scientific research. The coupling performance is well validated through comparison with experimental results, and CFD simulations of TKE distribution in the regulating well show that turbulence is more intense in the low-water-level transition condition.

Stable operation of transitions in the regulated plants can effectively reduce wind and solar abandonment rates and enhance their power acceptance capacity, and it further contributes to the construction of new energy power grids. WHED's application to investigate transient characteristics of more PSPs and HPs and analyze the impact of unstable transitions on the grid will be the next study for the authors.

Author Contributions: X.M.: Conceptualization, Formal analysis, Funding acquisition, Investigation, Methodology, Project administration, Supervision, Writing—review and editing. G.W.: Data curation, Formal analysis, Investigation, Software, Validation, Writing—original draft. Y.W.: Methodology, Resources. J.H.: Writing—original draft. X.G.: Writing—review and editing. P.Z.: Software. All authors have read and agreed to the published version of the manuscript.

Funding: This work is supported by the National Science Foundation of China (51909222), the Postdoctoral Special Funding Project of Shaanxi Province (2023BSHTBZZ22), The second "Young Talent Promotion" held by China Society for Hydropower Engineering (CSHE-YESS-2024006), China Postdoctoral Science Foundation under Grant 2024M752625, Third-class Postdoctoral Grant Program of Shaanxi Province (2024BSHSDZZ234).

Data Availability Statement: The data presented in this study are available on request from the corresponding author due to project requirements.

Conflicts of Interest: The authors declare that they have no known competing financial interests or personal relationships that could have appeared to influence the work reported in this paper.

Abbreviations and Symbols

The following abbreviations and symbols are used in this manuscript:

Abbreviations	
WDS	Water diversion system
WHED	Water hammer energy difference
PSP	Pumped storage plant
HP	Hydropower plant
MOC	Method of characteristics
1D	One-dimensional
3D	Three-dimensional
MDS	Multidimensional scale
EXP	Experiment
SSR	Sum of squares of residuals
TSS	Total sum of squares
Symbols	
v	Flow velocity of pipe cross-section, m/s
h	Water head, m
l	Length of pipe, m
d	Diameter of pipe, m
a	Wave velocity of water hammer, m/s
λ	Head loss coefficient, -
H_0	Initial water head, m
v_0	Initial flow velocity, m/s
E	Transfer parameters of positive energy, -
e	Transfer parameters of negative energy, -
A	Sectional area of pipe, m^3
n	Rotation speed, r/min
M	Torque, kN·m
Q	Flow rate, m^3/s
Q_{11}	Unit flow rate, m^3/s
Q_r	Rated flow rate, m^3/s
H_r	Rated water head, m
H_{11}	Unit water head, m
n_r	Rated rotation speed, r/min
n_{11}	Unit rotation speed, r/min
M_r	Rated torque, kN·m
t	Time, s
y	Relative opening of guide vane, -
y_v	Relative opening of guide valve, -
H_R	Reservoir level, m
R^2	Linearly dependent coefficient, -
H_w	Water level of surge tank, m
H_B	Bottom pressure of surge tank, m
Δ	Maximum fault tolerance, -
u_1	Amplitude of parameter, -
u_2	Change rate of parameter, -
u	Stability coefficient, -
$\overline{u}$	The weighted average, -
c	Sum of $\overline{u}$, -
δ_c	Velocity scale
δ_l	Geometric scale
Ω	Differences between predicted and calculated values, -
WH, *WM*	Flow rate and torque coefficient after Suter transformation, -

References

1. Liu, X.; Ni, Y.; Sun, Y.; Wang, J.; Wang, R.; Sun, Q. Multi-VPPs power-carbon joint trading optimization considering low-carbon operation mode. *J. Energy Storage* **2024**, *83*, 110786.
2. Ma, X.; Zhai, Y.; Zhang, T.; Yao, X.; Hong, J. What changes can solar and wind power bring to the electrification of China compared with coal electricity: From a cost-oriented life cycle impact perspective. *Energy Convers. Manag.* **2023**, *289*, 117162. [CrossRef]
3. Morabito, A.; Hendrick, P. Pump as turbine applied to micro energy storage and smart water grids: A case study. *Appl. Energy* **2019**, *241*, 567–597.
4. Ali, M.; Ahmad, M.; Koondhar, M.A.; Akram, M.S.; Verma, A.; Khan, B. Maximum power point tracking for grid-connected photovoltaic system using adaptive fuzzy logic controller. *Comput. Electr. Eng.* **2023**, *110*, 108879. [CrossRef]
5. Oskouei, M.Z.; Yazdankhah, A.S. Scenario-based stochastic optimal operation of wind, photovoltaic, pump-storage hybrid system in frequency-based pricing. *Energy Convers. Manag.* **2015**, *105*, 1105–1114. [CrossRef]

6. Fang, G.; Chen, G.; Yang, K.; Yin, W.; Tian, L. How does green fiscal expenditure promote green total factor energy efficiency?-Evidence from Chinese 254 cities. *Appl. Energy* **2024**, *353*, 122098. [CrossRef]
7. Luo, X.Q.; Zhu, G.J.; Feng, J.J. The technology progress and development trend of hydraulic turbine. *J. Hydroelectr. Eng.* **2020**, *39*, 1–18. (In Chinese)
8. Chen, Y.; Deng, C.; Wu, H.; Li, D.; Chen, M.; Peng, P.; Zhao, Y. Multi-stage Soft Coordinated Control of Variable Speed Pumped Storage Unit in the Process of Mode Conversion Under the Generation Condition. *Proc. CSEE* **2021**, *41*, 5258–5274. (In Chinese)
9. Hu, J.; Yang, J.; He, X.; Zhao, Z.; Yang, J. Transient analysis of a hydropower plant with a super-long headrace tunnel during load acceptance: Instability mechanism and measurement verification. *Energy* **2023**, *263*, 125671. [CrossRef]
10. Zhao, H.J.; Liu, S.B.; Zhang, Y.H. Study on reservoir ice regime of Pushihe pumped-storage power station. *Water Power* **2019**, *45*, 90–94. (In Chinese)
11. Yuan, B.; Li, C.; Feng, Y. Partition of operating regions in turbine mode for No.4 unit of Pushihe pumped storage power station. *Water Resour. Power* **2017**, *35*, 153–157. (In Chinese)
12. Lu, J.; Wu, G.; Zhou, L.; Wu, J. Finite Volume method for modeling the load-rejection process of a hydropower plant with an air cushion surge chamber. *Water* **2023**, *15*, 682. [CrossRef]
13. Ma, Z.; Zhu, B. Pressure fluctuations in vaneless space of pump-turbines with large blade lean runners in the S-shaped region. *Renew. Energy* **2020**, *153*, 1283–1295. [CrossRef]
14. Nasir, J.; Javed, A.; Ali, M.; Ullah, K.; Kazmi, S.A.A. Capacity optimization of pumped storage hydropower and its impact on an integrated conventional hydropower plant operation. *Appl. Energy* **2022**, *323*, 119651. [CrossRef]
15. Yang, Z.; Liu, Z.; Cheng, Y.; Zhang, X.; Liu, K.; Xia, L. Differences of Flow Patterns and Pressure Pulsations in Four Prototype Pump-Turbines during Runaway Transient Processes. *Energies* **2020**, *13*, 5296. [CrossRef]
16. Li, Z.; Bi, H.; Karney, B.; Wang, Z.; Yao, Z. Three-dimensional transient simulation of a prototype pump-turbine during normal turbine shutdown. *J. Hydraul. Res.* **2017**, *55*, 520–537. [CrossRef]
17. Su, W.-T.; Li, X.-B.; Xia, Y.-X.; Liu, Q.-Z.; Binama, M.; Zhang, Y.-N. Pressure fluctuation characteristics of a model pump-turbine during runaway transient. *Renew. Energy* **2021**, *163*, 517–529. [CrossRef]
18. Henclik, S. Analytical solution and numerical study on water hammer in a pipeline closed with an elastically attached valve. *J. Sound Vib.* **2018**, *417*, 245–259. [CrossRef]
19. Urbanowicz, K.; Bergant, A.; Stosiak, M.; Deptuła, A.; Karpenko, M.; Kubrak, M.; Kodura, A. Water Hammer Simulation Using Simplified Convolution-Based Unsteady Friction Model. *Water* **2022**, *14*, 3151. [CrossRef]
20. Yang, Z.; Cheng, Y.; Xia, L.; Meng, W.; Liu, K.; Zhang, X. Evolutions of flow patterns and pressure fluctuations in a prototype pump-turbine during the runaway transient process after pump-trip. *Renew. Energy* **2020**, *152*, 149–1159. [CrossRef]
21. Hou, J.; Li, G.H.; Li, X.X.; Wu, X.M.; Chen, Y.M. *Transient Process of Complex Hydraulic System*; China & Water Power Press: Beijing, China, 2019. (In Chinese)
22. Zhang, L.; Wu, Q.; Ma, Z.; Wang, X. Transient vibration analysis of unit-plant structure for hydropower station in sudden load increasing process. *Mech. Syst. Signal Process.* **2019**, *120*, 486–504. [CrossRef]
23. Telikani, A.; Rossi, M.; Khajehali, N.; Renzi, M. Pumps-as-Turbines (PaTs) performance prediction improvement using evolutionary artificial neural networks. *Appl. Energy* **2023**, *330*, 120316. [CrossRef]
24. Zhang, Y.; Xu, Y.; Zheng, Y.; Rodriguez, E.F.; Liu, H.; Feng, J. Analysis on Guide Vane Closure Schemes of High-Head Pumped Storage Unit during Pump Outage Condition. *Math. Probl. Eng.* **2019**, *2019*, 8262074. [CrossRef]
25. Yu, X.; Zhang, J.; Zhou, L. Hydraulic Transients in the Long Diversion-Type Hydropower Station with a Complex Differential Surge Tank. *Sci. World J.* **2014**, *2014*, 241868. [CrossRef] [PubMed]
26. Li, H.; Wan, Z.; Huang, W.; Zeng, M.; Fang, J.; Xie, J. Robust comprehensive evaluation of guide vane closure law in hydraulic turbines in moderately-high head hydropower plants. *J. Tsinghua Univ. (Sci. Technol.)* **2023**, *63*, 125–133. (In Chinese)
27. Wang, Y.; Zhang, J.; Xu, T.; Liu, Y.; Yao, T.; Wang, K.; Zhang, M. Air valve arrangement criteria for preventing secondary pipe bursts in long-distance gravitational water supply systems. *Aqua-Water Infrastruct. Ecosyst. Soc.* **2023**, *72*, 1566–1581. [CrossRef]
28. Trivedi, C.; Gandhi, B.K.; Cervantes, M.J.; Dahlhaug, O.G. Experimental investigations of a model Francis turbine during shutdown at synchronous speed. *Renew. Energy* **2015**, *83*, 828–836. [CrossRef]
29. Ye, J.; Do, N.C.; Zeng, W.; Lambert, M. Physics-informed neural networks for hydraulic transient analysis in pipeline systems. *Water Res.* **2022**, *211*, 118828. [CrossRef]
30. Chen, S.; Wang, J.; Zhang, J.; Yu, X.; He, W. Transient behavior of two-stage load rejection for multiple units system in pumped storage plants. *Renew. Energy* **2020**, *169*, 1012–1022. [CrossRef]
31. Mao, X.; Zhong, P.; Wang, Y.; Tan, Q.; Cui, Q. Study on the evolution of transient flow field in WDS based on numerical and experimental methods. *J. Energy Storage* **2023**, *73*, 108996. [CrossRef]
32. Zhang, W.; Deng, J.; Wang, P.; Wang, Y. Study on similitude method for turbine considering working fluid physical properties variation. *Appl. Energy* **2023**, *338*, 120930. [CrossRef]

33. Zhang, P.; Cheng, Y.; Xue, S.; Hu, Z.; Tang, M.; Chen, X. 1D-3D coupled simulation method of hydraulic transients in ultra-long hydraulic systems based on OpenFOAM. *Eng. Appl. Comput. Fluid Mech.* **2023**, *17*, 2229889. [CrossRef]
34. Bi, H.; Chen, F.; Wang, C.; Wang, Z.; Fan, H.; Luo, Y. Analysis of dynamic performance in a pump-turbine during the successive load rejection. *IOP Conf. Ser. Earth Environ. Sci.* **2021**, *774*, 012152. [CrossRef]
35. Cherny, S.; Chirkov, D.; Bannikov, D.; Lapin, V.; Skorospelov, V.; Eshkunova, I.; Avdushenko, A. 3D numerical simulation of transient processes in hydraulic turbines. *IOP Conf. Ser. Earth Environ. Sci.* **2010**, *12*, 012071. [CrossRef]
36. Zhang, X.-X.; Cheng, Y.-G.; Yang, J.-D.; Xia, L.-S.; Lai, X. Simulation of the load rejection transient process of a francis turbine by using a 1D-3D coupling approach. *J. Hydrodyn.* **2014**, *26*, 715–724. [CrossRef]
37. Cheng, Y.C.; Zhang, J.B.; Chen, G.; Zhang, D. *Automatic Regulation of Water Turbine*; China Water & Power Press: Beijing, China, 2010. (In Chinese)
38. Zheng, Y.; Zhang, J. *Hydro Unit Transition Process*; Peking University Press: Beijing, China, 2008. (In Chinese)
39. Mao, X.L.; Wen, G.Q. *Pumped Storage Power Station Power Generation Mode Full Load Condition and Pumping Mode Power Off Condition Transition Process Calculation Software*, version 2023SR1542053; Northwest A & F University: Xianyang, China, 2023. (In Chinese)

energies

MDPI

Article

Failure Modes and Effect Analysis of Turbine Units of Pumped Hydro-Energy Storage Systems

Georgi Todorov [1], Ivan Kralov [2], Konstantin Kamberov [1], Yavor Sofronov [1], Blagovest Zlatev [1] and Evtim Zahariev [1,*]

1 Faculty of Industrial Technology, Technical University of Sofia, 1000 Sofia, Bulgaria; gdt@tu-sofia.bg (G.T.); kkamberov@tu-sofia.bg (K.K.); ysofronov@tu-sofia.bg (Y.S.); bzlatev@tu-sofia.bg (B.Z.)
2 Faculty of Transport, Technical University of Sofia, 1000 Sofia, Bulgaria; kralov@tu-sofia.bg
* Correspondence: evtimvz@gmail.com

Abstract: In the present paper, the subject of investigation is the reliability assessment of the single-stage reversible Hydropower Unit No. 3 (HU3) in the Bulgarian Pumped Hydro-Electric Storage (PHES) plant "Chaira", which processes the waters of the "Belmeken" dam and "Chaira" dam. Preceding the destruction of HU4 and its virtual simulation, an analysis and its conclusions for rehabilitation and safety provided the information required for the reliability assessment of HU3. Detailed analysis of the consequences of the prolonged use of HU3 was carried out. The Supervisory Control and Data Acquisition (SCADA) system records were studied. Fault Tree Analysis (FTA) was applied to determine the component relationships and subsystem failures that can lead to an undesired primary event. A Failure Modes and Effect Analysis methodology was proposed for the large-scale hydraulic units and PHES. Based on the data of the virtual simulation and the investigations of the HU4 and its damages, as well as on the failures in the stay vanes of HU3, it is recommended to organize the monitoring of crucial elements of the structure and of water ingress into the drainage holes, which will allow for detecting failures in a timely manner.

Keywords: FMEA; accident analysis; hydro-energy; pumped hydro-energy storage; fault tree analysis; failure prediction

1. Introduction

This article is related to the contract of the "National Electric Company" EAD (NEK AD) of Bulgaria [1] for the investigation and analysis of the possibility of the safe operation of HU3 at the "Chaira" PHES plant. A brief history of the construction and operation of the Chaira HPES can be found in [2].

The objective of the contract was the investigation of the reliability of the single-stage reversible HU3 at the "Chaira" PHES plant, which processes the waters of the "Belmeken" dam and the "Chaira" dam (Figure 1). The "Belmeken" dam is located at an altitude of about 2000 m. The upper water level of the dam is at an altitude of 1920 m. The lowest level is at 1865 m. Figure 1 also shows the altitude of the dam walls. The dam contains 144 million cubic meters of water. The "Chaira" fam is located at an altitude of about 1200 m. It contains about 5 million cubic meters of water. The upper water level of the dam is 1260 m, and the lower level is 1231 m. The plant consists of four hydro units (HUs) that are located at an altitude of 1169 m. The power generated by the "Chaira" PHES is 864 megawatts, and in pumping mode it consumes 788 megawatts. The water pipes are divided into two levels, each with two pipes with diameters of 4.4–4.2 m. The level of the upper section is at an altitude of about 1900 m. The total length of the pipelines is 2510 m for the upper level and 1816 m for the lower level, respectively. The flow rate of each pipe is about 7.2 cubic meters per second.

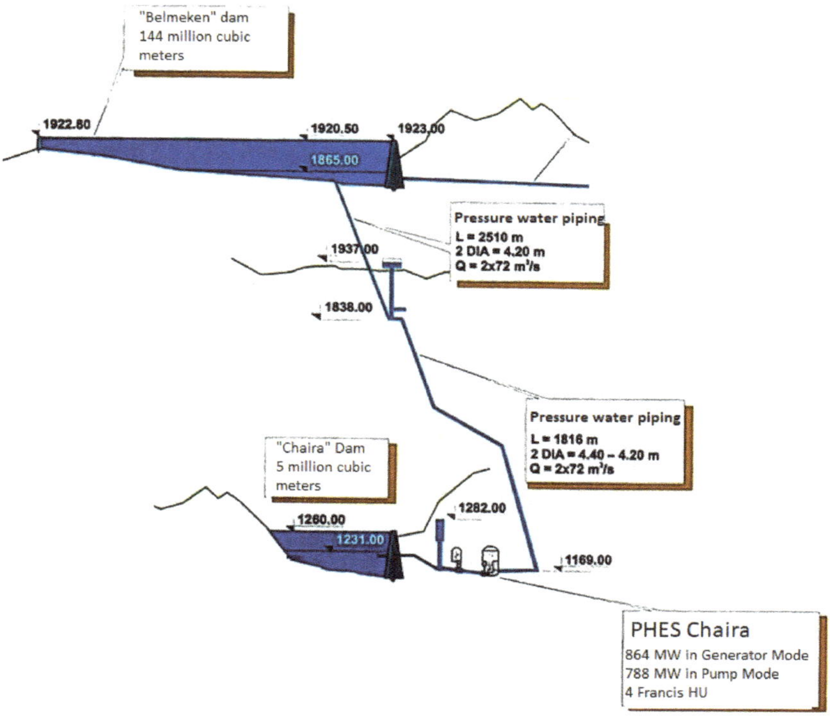

Figure 1. The principal scheme of the PHES "Chaira" plant.

The scheme of each powerhouse is presented in Figure 2. The high-pressure water flow enters from the "Belmeken" dam (Figure 1) into each HU (Figure 2) through pressure water pipe 6 (penstock) in powerhouse 1 and drives the turbine 1b and the motor–generator in generator mode. In pump mode, pump 1b is turned on, which is driven by the motor–generator in engine mode. The waste water flows through pipeline 7 (tailraces) into the lower reservoir "Chaira" (Figure 1). The water flow is interrupted through the downstream gates 7a. Transformers and power lines are installed in cable gallery 8, and part of the water flow is accumulated in lower surge shafts 3 to equalize the pressure and prevent water hammer. The water flow is regulated through the downstream gates 7a.

The rehabilitation of the HU3 unit began on 20 February 2021 with the completion of repair and installation activities and the adjustment of the unit's systems. As a result of the damages that occurred in HU4, on 2 June 2022, and before the start of the trial testing stages of HU3, the activities were terminated by the contractor in order to prevent similar damage as that of HU4 [3].

Cases of failures in hydroelectric power plants and, more specifically, in HPESs are not unusual worldwide. Such cases have been described in the literature since the beginning of the 20th century. In [4], many incidents are listed, although it is not a complete list. Predominantly Francis turbines are applied either as turbines or as pumps, since this type of structure is especially effective for both cases. Many cases of Francis turbine failures for the period from 1990 to 2010 are analyzed in the paper of Yasuda and Watanabe [5]. Incidents were reported for power plants in Canada, China, Australia, Iran, Nepal and the

USA. One of the most dramatic cases was reported for the Sayano-Shushenskaya power station [6]. In the scientific literature, the destructions and failures of the turbine blade are mainly discussed. In [5], there are no cases analyzed regarding the damages of the stay vanes. Severe vibrations were the reasons for the cracks found in the runners of Francis turbines [7]. Damages as a result of the erosion of the turbine blades and guide vanes were reported in Nepal in 2003 [8]. Cracks were found at the trailing edges of the turbine blades of Francis turbines in Iran in 2006 [9] and in Canada in 2008.

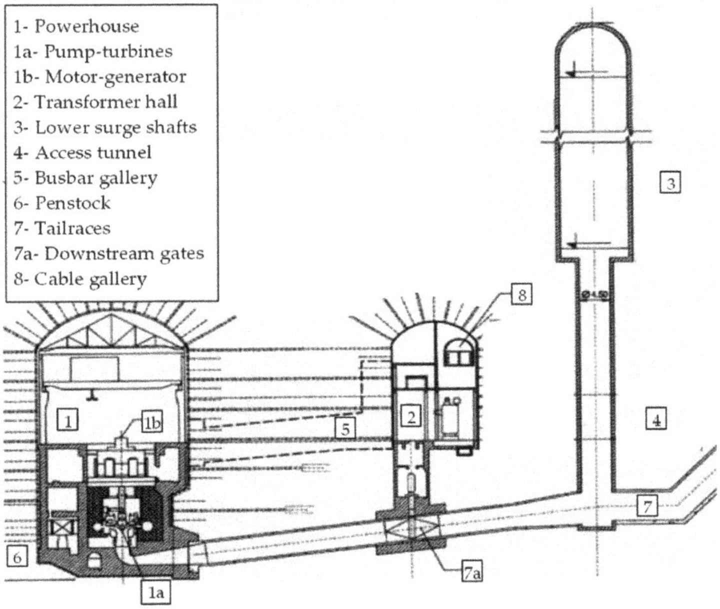

Figure 2. The structure of a unit of PHES at the "Chaira" plant.

Another case of the destruction of a Francis turbine of 88 MW was reported in the USA in 2008. The runner blades had severe damage due to hard-hitting by a freed guide vane. A similar case was reported in the Tocantins River in central Brazil [10]. The cause was the dropping of a link pin of the guide vane operating mechanism. In 2010, in Canada [11], many cracks were found at the flange fillet of the main shaft at the runner side.

Some cases of spiral casing destructions of Francis turbines are also discussed. In Australia, in 1990, a Francis turbine of 150 MW experienced spiral case failure due to an excessive pressure increase due to the instant shutdown of all guide vanes [12]. Spiral case embedment and destruction was discussed in several articles by Chinese scientists [13,14].

It should be noted that investigations of possible destructions of Francis turbine stay vanes could not be found in the scientific literature. Todorov et al. [15–17] analyzed the destruction of the Francis turbine stay vanes of the PHES "Chaira" HU4 and the possible reasons for the occurrence of cracks. They conducted detailed investigations on the influence of concrete erosion [15] on the destruction of the spiral casing and the stay vanes [16]. The fatigue of the material was also discussed [16,17].

The unprecedented accident at HU4 of the Chaira HPP [17] necessitated the termination of the rehabilitation of HU3. The main cause of the accident at HU4 was the complete destruction of all blades, an event that has not been described in scientific publications. In Figure 3 the stay vane No. 1 of HU4 and the cracks are shown. It was proven as a result of the virtual simulations in [16,17] that the reason for the stay vanes' destruction was the low-cycle fatigue of the material. The red circle shows the stay vane No. 1, where the yellow sector is the part with inadmissible stresses and strains. These are the places where the cracks appeared. Surveying the scientific literature showed that the problems of the fatigue life and the service live of Francis turbines were within the scope of many publications. In [18], it is recommended that fatigue safety factors to be more than 1.5 and guidelines are proposed for the determination of fatigue cycles and crack propagation calculations. In [19], the fatigue reliability of welded steel structures is analyzed. Liu et al. [20] reviewed the publications on the fatigue damage mechanisms in hydro turbines. Lyutov et al. [21] used the stress pulsation amplitude to estimate the number of cycles until the moment of fatigue failure. The numbers of loading cycles and oscillation frequency are also used to calculate the runner service time. Paresas et al. [22] estimated the fatigue lives of Francis turbines based on experimental strain measurements. Biner et al. [23] performed a numerical fatigue damage analysis of a variable-speed Francis pump turbine in start-up and generating modes. In case of variable loading conditions, the use of the correct factor of safety in structural strength calculations is of particular importance. Zhang et al. [24] studied the major factors affecting the fatigue life prediction of steel spiral cases in pumped-storage power plants. They expected that the factors identified in the paper would assist in understanding the role of adequate fatigue design and the analysis of PHES plants.

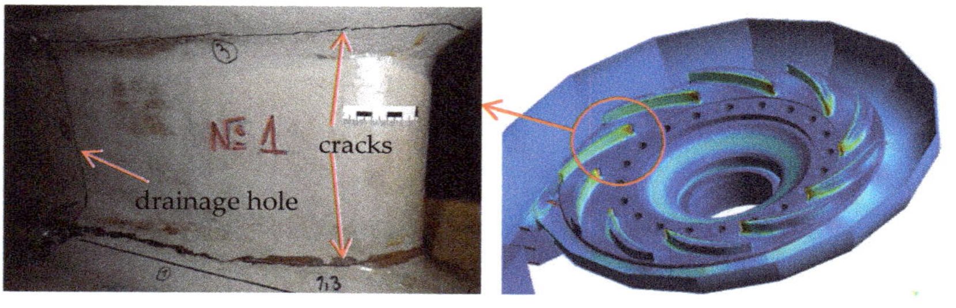

Figure 3. The cracks of stay vane No. 1 of HU4 [17].

A key element in the analysis of failures and, most importantly, the probability of their occurrence is the risk assessment methodology. A type of methodology that is increasingly being used in modern products is the Failure Modes, Effects and Analysis (FMEA) methodology. Its premise is the availability of quantitative estimates of the probability of a given type of failure occurring, and it examines in greater detail the types of failure of the facility.

A suitable method for conducting reliability and safety analyses is Fault Tree Analysis (FTA). This method uses systems analysis to determine the component relationships and subsystem failures that can lead to an undesired event, known as a primary event. The automotive industry mainly uses FMEA, while the aerospace industry uses FTA. In many cases, the best results are obtained by combining several analysis methods.

Souza and Álvares [25] applied FMEA and FTA for the assessment of the reliability of hydraulic Kaplan turbines used in the hydroelectric plant of Balbina, Amazonas, Brazil. They showed the contribution of each one to predictive maintenance. Peeters et al. [26] assessed the FMEA model in order to select the critical system-level failure modes. For each of them, a function-level FTA was performed, followed by an FMEA. The Infraspeak Team [27] published a paper about the differences between the FTA and FMEA models. It was shown that each analysis has its own approach to failures, which could lead to different results.

Another type of analysis included in some international standards does not require a quantitative assessment of the probability of a particular type of failure but only a description of the possible failures, their effects and the risk of failure (criticality). This type of analysis is known as risk hazard analysis (RHA). This type of analysis is described in detail by the standards DIN EN ISO 12100 [28].

Flynn [29] discussed the methods used to identify hazards and the causes and consequences of accidents. It was emphasized that many accidents occurred because of a lack of knowledge of the system, process or substance being dealt with. TheSafeyMaster published a study report [30] for consultancy and training services about hazard identification and risk assessment. It was clearly stated that hazard identification and risk assessment are critical processes that organizations need to undertake to ensure the safety of their employees.

This article examined the possibility of accidents occurring at HU3 of the PHES "Chaira" plant. The causes of the HU4 accident were analyzed and the design documentation, the installation process, the SCADA system records, the protocols of the current repairs performed, and the results of the virtual modeling of the hydro units were studied. The results of the HU4 investigations were compared with the data from the investigations of HU3, which did not suffer an accident, as it is identical to HU4, both in terms of construction and installation performance. The results of the SCADA system records and information, as well as the operation data of HU3, were compared with those of the damaged HU4.

The probability of system failure was estimated based on the failure probabilities of the primary events. The creation of the so-called fault tree was based on system and functional analysis, the definition of the unwanted event (failure at the basic level), the determination of the types and categories of failures, the depiction of the effectiveness of failures in the fault tree to the main events, the assessment of the main events from the input data (failure frequency, duration of the events) and the probabilistic assessment of the fault tree. The effects of static loads, dynamic loads and low-cycle loads were investigated. A systematic risk analysis was carried out. The results of the application of FMEA and FTA were supplemented with the results of the hazard and operability study of RHA, with the best solutions being achieved through a combination of the analysis methods. Since the existing methods for risk assessment are mainly used for the aeronautics and automobile industries, the innovation of the proposed methodology consisted of improving it applicability to large-scale hydraulic units such as PHES.

Based on experience and the investigations of the HU4 and its damages, as well as of some failures in the stay vanes of HU3, it is recommended, according to Section 5, to conduct monitoring of the major parts of the HUs, as well as for water ingress into the drainage holes to be organized. This will allow for detecting failures in a timely manner.

2. Materials and Methods

2.1. Risk Analysis: Essence

Risk hazard analysis (RHA), set out in some international standards and safety requirements for a number of devices, does not require a quantitative assessment of the probability of given types of failures. Only a description of the possible failures, their effects and the risk of failure occurring (criticality) is needed. It is used to identify and assess potential risks in the use of a device. The measures taken to ensure operational safety are to be documented, and it is not required to include all measures taken with regard to the safety of the device. It should be noted that technical and formal errors are possible when preparing the documentation.

The presented description of RHA is based on and described in detail by the German standard DIN EN ISO 12100 Part1 [28]. It applies to lifting facilities but also is used for other equipment. The specific possible risks and their causes are the main objects of consideration. The analysis report should also contain prescriptions for operational control based on knowledge of the nature of the occurrence of failures and their specificities. The identification of the risk level is obtained through the assessment of the risk and the corresponding types of dangerous failures. The frequencies and the possibility of failure occurrences are determined according to a scale defined in the standard and shown in Figure 4.

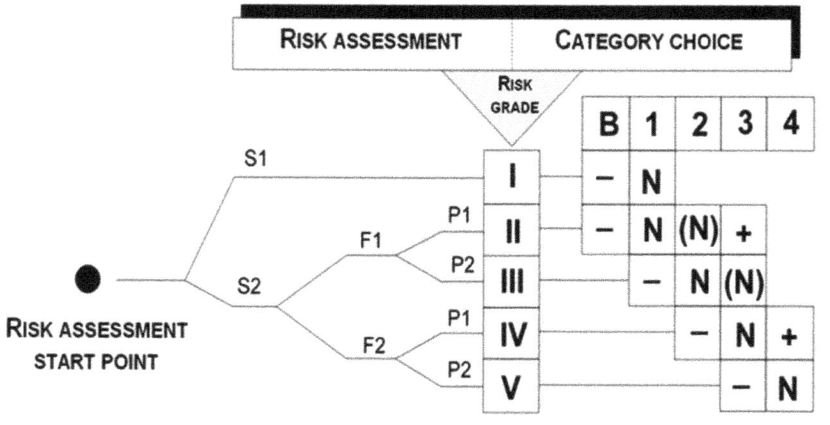

Figure 4. The risk assessment system for performing RHA according to DIN EN ISO 12100 Part1.

The notations presented on the Figure 4 are as follows:

B: 1–4: Categories for security-related parts of controls;
N: Normal category for the risk level
(N): Additional directions for standard solutions for protection devices and electronical devices (this category can generally be accomplished by using electronics);
+: Deviation to an upper category;
−: Deviation to a lower category.

Hazard analysis documentation is prepared in order to make the hazard more clear and understandable, and, thus, the hazardous location, the hazard cause and the operating method are also listed. The definition of the type of hazard is evaluated as follows:

(S): Grade of the possible injury:

S1: Light injury;
S2: Severe permanent injury.

(F): Frequency of the incidence:

F1: Rarely to often;
F2: Frequent to always.

(P): Possibility of risk prevention:

P1: Possible risk prevention under certain circumstances;
P2: Almost impossible.

The DIN EN ISO 14121-1 [31] standard defines risk as a combination of the probability of damage occurring and its degree of criticality. There are a large number of procedures for analyzing these factors. In general, two main types of risk analysis are applied—deductive and inductive. The deductive procedure starts with an event and analyzes its causes. The inductive procedure assumes the existence of possible deviations in a process or a system and analyzes their effects.

The technical context of the present study requires that the concept of safety analysis be considered, although the term risk analysis is often used in connection with economic analyses.

2.2. Methodology Used for the Analysis of Failures and Their Effects

The main method used for reliability analysis and the definition of the probability of damage and destruction is based on the investigation performed via Failure Modes, Effects and Criticality Analysis (FMECA) [32]. This method is increasingly being used in modern products. The premise is the availability of quantitative estimates of the probability of a given type of failure and its detailed examination. The FMECA is performed prior to any failure actually occurring. FMECA analyzes risk, which is measured by criticality (the combination of severity and probability), to take action and, thus, provide an opportunity to reduce the possibility of failure.

FMECA and Failure Mode and Effects Analysis (FMEA) [25–27] are closely related tools. There are two activities used to perform FMECA: creating FMEA; performing the criticality analysis. Each tool resolves to identify failure modes that may potentially cause product or process failure. FMEA is qualitative, exploring "what-if scenarios", where FMECA includes a degree of quantitative input taken from a source of known failure rates. A source for such data is Military Handbook 217 [33] or an equivalent source.

As already mentioned, there are a large number of methods for performing analysis and evaluation. The automotive industry mainly uses FMEA, while the aerospace industry uses Fault Tree Analysis (FTA), although very often these two methods are applied sequentially [27]. The chemical industry often uses the hazard and operability (HAZOP) study methodology [34]. In many cases, the best results are obtained by combining several methods for risk and safety analysis.

The FMEA methodology was developed in the NASA space program in 1959/60. FMEA is applied to the study of potential weaknesses in the planning and design phase. This analytical methodology is of a preventive nature.

The analysis of certain risks involves considering each system unit and its association with the probability of hazard. An important element of FMEA is the determination of a quantitative expression of the risk, the risk priority number (RPN), which assesses the criticality of the specific failure. The RPN is determined as follows:

$$S \times O \times D = RPN, \tag{1}$$

where the following is true:

- S—severity (criticality)—assesses the degree of significance of a failure;
- O—occurrence (failure intensity)—assesses the likely occurrence of such a failure;
- D—detection (detectability)—represents the probability of detecting the cause of a failure.

The RPN value is used for decisions regarding the need for intervention and changes. The values indicate the following:

- RPN values up to 40 indicate low risk (no need for corrective actions);
- RPN values in the range $40 \div 100$ indicate moderate risk (certain actions are needed to improve the study object);
- RPN values above 100 are classified as unacceptable risk (urgent actions are needed).

The generally accepted values and descriptions of these parameters are given in Tables 1–3.

Table 1. Severity (S): criticality levels and their assessment.

Level	Description	Rating (S)
None	No effect on components	1
Minor	Minor effect on the system	2
Very low	Slightly pronounced impact on the system	3
Low	Low level of criticality regarding the functioning of the system	4
Average	The system is functioning, albeit with broken parameters	5
High	Reduced system functionality	6
Very high	Loss of important system functions	7
Dangerous	Functions are lost, leading to potential danger to users	8
Very dangerous	Potentially dangerous system condition, with indications allowing preventive action	9
Extremely dangerous	System condition with possible critical impacts on personnel, albeit without possibility of detection and prevention	10

Table 2. Occurrence (O): failure intensity assessment.

Intensity	Probability	Rating (O)
Extremely low	$\leq 1 \times 10^{-5}$	1
Low	1×10^{-4}	2
Average grade	5×10^{-4}	3
	1×10^{-3}	4
	2×10^{-3}	5
High degree (repeatability)	5×10^{-3}	6
	1×10^{-2}	7
	2×10^{-2}	8
	5×10^{-2}	9
Very high degree	$\geq 1 \times 10^{-1}$	10

Table 3. Detection (D): detection rate.

Grade	Description	Rating (D)
Very high	Very high probability of failure detection	1
High	High probability of failure detection	2
Relatively high	Relatively high probability of failure detection	3
Medium	Average probability of detecting failure	4
Relatively low	Relatively low probability of detecting the potential cause/mechanism of failure	5

Table 3. *Cont.*

Grade	Description	Rating (D)
Low	Low probability of detecting the potential cause/mechanism of failure	6
Very low	Very low probability of detecting the potential cause/mechanism of failure	7
Weak	Weak probability of detecting the potential cause/mechanism of failure	8
Very weak	Very weak probability of detecting the potential cause/mechanism of failure	9
Impossible	Inability to establish the failure	10

3. Results

Fault Tree Analysis of HU3 of PHES "Chaira"

FTA is suitable for conducting reliability and safety analyses. The methodology uses system analysis to determine the relationships and subsystem failures that could lead to an undesired event, known as a primary event. The FTA enables the representation of the functional structure of the system as a causal chain of failures and their effects. The main aim is to estimate the probability of total system failure based on the probabilities of the main failure events occurring. The FTA shows which failures cause emergency events, and the aim is evaluating and predicting possible preventive measures. Further, a quantitative analysis is performed to calculate the probability of the occurrence of an undesirable event.

In general, the following components and fault categories are used:

- Primary failure (failure of a component under normal operating conditions);
- Secondary failure (failure of a component as a result of secondary failure from a primary failure or as a result of extreme operating conditions);
- Errors as a result of incorrect operation or misuse.

An important factor is the nature of the fault linkage. While an "OR" combination of two inputs is sufficient to trigger a fault, an "AND" connection requires both inputs for it to occur.

The creation of the so-called fault tree occurs during the following stages: performing system and functional analysis; defining the unwanted event (failure at the basic level); determining the types and categories of failures; depicting the effectiveness of failures in the fault tree to the main events; evaluating the main events from input data (failure frequency, times); performing probabilistic assessment of the fault tree (calculation of the above event).

A section of the Francis turbine and its simplified mayor units and parts are shown in Figure 5. The following parts are denoted by numbers: 1—the concrete surrounding the structure; 2—the spiral casing; 3—the stay vanes; 4—the guide vanes; 5—the bolts of the upper 6 and the lower 7 covers; 8—the bearing; 9—the runner. The corresponding units are in different colors.

A simple example of a possible failure during HU operation is shown. As a result of increasing the gap between spiral casing 2 and concrete 1, the loading and the deflections in stay vanes 3 become unacceptable and cracking appears. Then, the destruction of stay vanes 3 follows and the destruction of bolts 5 connecting the upper 6 and lower 7 covers follows. Water penetrating through bearings 8 appears. This process was observed during the failure of HU4.

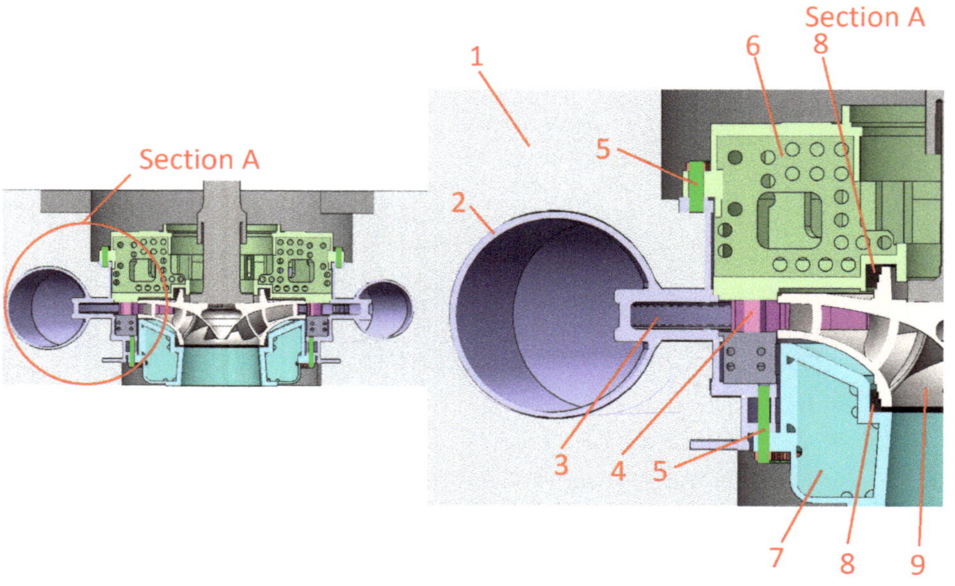

Figure 5. A section of the Francis turbine and its main units and details.

Based on the results, it is possible to determine the most effective measures to eliminate weak points and optimize reliability and safety. This analysis is related to the possible failures of the stay vanes of the spiral casing (stator columns) and the effects caused by them. The following possible failures of the stay vanes are described further down.

- Primary failures/shutting out (PFs):
  - PF1: Crack formation on the faces of up to three stay vanes due to low-cycle material fatigue;
  - PF2: Crack formation on the faces of more than three stay vanes due to low-cycle material fatigue;
  - PF3: The violation of the bond between concrete and spiral casing, leading to a backlash.
- Secondary failures/shutting out (SFs):
  - SF1: The failure of up to three stay vanes;
  - SF2: The failure of all stay vanes;
  - SF3: Significant deformations in the spiral casing;
  - SF4: Increased load on the lower and upper covers;
  - SF5: Increased load on the bolted connections of the covers due to their overloading by bending moment.
- Effects of failures (EFFs):
  - EFF1: Deteriorated guide vane bearings—the violation of clearance and coaxiality occurs between the guide vanes and the bearings, leading to difficulty closing (switching off) the vane control;
  - EFF2: Deteriorated runner to spiral casing clearance—the violation of clearance between the runner and the spiral casing and possible mutual contact;
  - EFF3: Damaged bolted connections of covers—the destruction of bolted connections of the covers due to their overloading by bending moment;

- EFF4: Cracks in the concrete—the cracking of the concrete due to the overloading of the spiral casing and the total failure of the stay vanes.

These failures and their effects are used to construct the fault tree, which is drawn as a pictogram that highlights the system relations. It is shown in Figure 6.

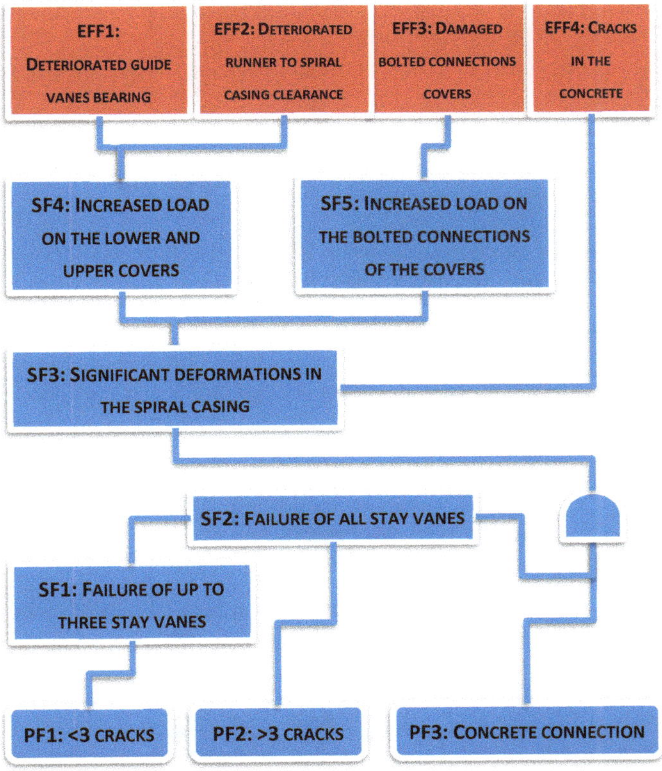

Figure 6. Fault tree related to stay vanes.

The possible failures modes are systematized using the compiled failure tree presented in Table 4.

Table 4. Possible failure modes and their effects.

Effect of Refusal	Mode	Effect	Mark
EO1: The violation of the clearance and alignment between the guide vanes and bearings, leading to difficult or no control	P1/P2/	Strong vibrations in the structure; Water appearing in the service area through drainage holes in the stay vanes.	EO1.1
	G1/G2	Strong vibrations in the structure; Water appearing in the service area through drainage holes in the stay vanes; Difficulty closing the water flow and switching off the machine	EO1.2
	A1	Strong vibrations in the structure; Water appearing in the service area through drainage holes in the stay vanes; A rapid increase in the machine rotation frequency and danger of exceeding critical ones, leading to destruction; Serious damage to the electrical part of the system	EO1.3

Table 4. *Cont.*

Effect of Refusal	Mode	Effect	Mark
EO2: The violation of the clearance between the runner and the spiral casing and contact	G1/G2/P1/P2/A1	Strong impacts on the structure; Water in the service area; The risk of the destruction of the runner bearing; Possible mechanical damage to the spiral casing and the runner	EO2
EO3: The destruction of bolted connections of the covers due to overloading by bending moment	G1/G2/P1/P2/A1	Strong impacts on the structure; Massive water ingress into the engine room; Difficult or impossible to close the guide vanes	EO3
EO4: The cracking of the concrete due to the overloading of the spiral casing and broken integrity of the stay vanes	G1/G2/P1/P2/A1	Severe deformations in the structure; Difficult or impossible to close the guide vanes Danger of the destruction of the runner bearing; Serious damages to the electrical part of the system	EO4

The failure modes and their effects are quantitatively assessed as the criticality level, intensity of occurrence and degree of detectability in Table 5.

Table 5. Evaluation of failure modes.

Failure Effect	Severity (S)	Occurence (O)	Detection (D)	RPN
EO1.1	2	2	6	24
EO1.2	3	2	6	36
EO1.3	8	2	2	32
EO2	6	2	1	12
EO3	10	1	2	20
EO4	9	2	1	18

Case EO4—the cracking of the concrete due to the overloading of the spiral casing and broken integrity of the stay vanes—can be found in Table 4. The operating modes are as follows: G1—nominal generator; G2—peak loads in the generator; P1—nominal pump mode; P2—peak loads in the pump; A1—drop of the load.

The parameter RPN (risk priority number) could be calculated according the classification in Table 5 from Equation (1): the risk priority number is calculated as $RPN = S \times O \times D = 18$, which could be counted directly from the matrix in Figure 7.

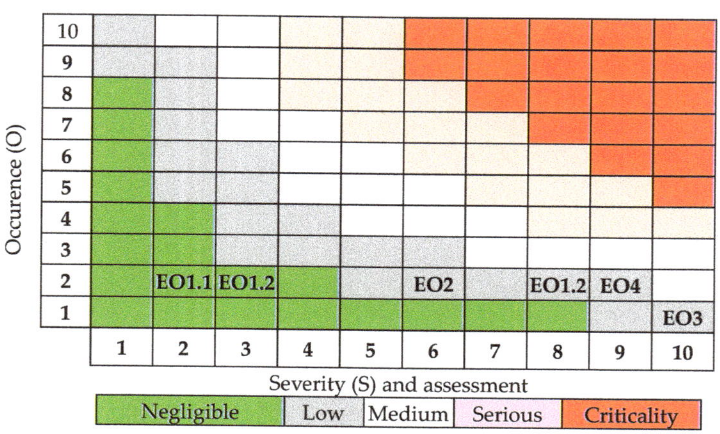

Figure 7. Failure criticality matrix.

4. Discussion

The processes of analysis related to the rehabilitation, damage and repair of the units of the HPES "Chaira" were carried out after the shutdown for rehabilitation of HA3.

Damages to the stator columns were found, which were not critical at this stage. Soon after that, a very serious damage occurred in HU4, which necessitated thorough analysis of both HU3 and HU4 in order to prevent critical accidents and to plan the appropriate repairs. After the detailed analysis of the failed PHES HU4 [16,17] and the analysis of the records of current repairs and emergency situations in HU3 and HU4, a number of recommendations can be made regarding the maintenance and control of the units in PHES "Chaira".

Particular attention should be paid to regular inspection and planned repairs. Events that depend on the degree of detectability should be taken into account. These are the units and places classified with 10 points for detectability D (inability to establish the failure, Table 3). Places and units that cannot be identified visually must be equipped with sensors and control devices. These are the guide vanes, the water ingress in the bearings, and the gap between the concrete and the spiral casing. Special attention should be paid to the deflections of the bolts of the upper and lower covers, as well as the stresses in the spiral casing.

Regular inspections of the stay vanes and the cavities on their surfaces because of the cavitation effects are needed. The welded parts of the stay vanes should be regularly monitored. The regimes for the welding of cavities and cracks should be not extreme and must not be the reason for changing the characteristics of the metal.

The results are also visualized as a criticality matrix, as shown in Figure 7.

5. Conclusions

The main task of this study was to propose measures and activities regarding the rehabilitation of HU3, PHES "Chaira", by analyzing the causes of the damage to its stator columns.

The analysis of the accident that occurred on the identical structure of HU4, PHES "Chaira", and full analysis of the concrete, spiral chamber, stator columns, loads, and strength and deformation characteristics of the materials of the critical elements are the basis of the comparative analysis of HU3, PHES "Chaira". The destruction of HU4 and its virtual simulation and analysis, as well as the conclusions for the rehabilitation processes and safety programs, provided information on the possible failure processes in HU3.

Detailed analysis of the consequences of prolonged use of HU3 was carried out. The records of the accidents and the rehabilitation processes were studied and used as the major information sources for the conclusions and proposal of the safety measures.

FTA was applied to determine the component relationships and subsystem failures that can lead to an undesired primary event.

The probability of system failure was estimated based on the failure probabilities of the primary events. The effects of static loads, dynamic loads and low-cycle loads were investigated.

Based on experience and the investigation of HU4 and its damages, as well as of the failures in the stay vanes of HU3, the following are recommended:

- Regular inspection and planned repairs be to provided.
- Units that cannot be surveyed visually must be equipped with sensors and control devices; these are the following:
    - The guide vanes and their welding places;
    - The water ingress in the bearings;
    - The gap between the concrete and the spiral casing;
    - The deflections of the bolts of the upper and lower covers;

  - The stresses in the spiral casing.
- Monitoring for water ingress into the drainage holes should be organized.
  These measures and additional equipment will allow for the timely detection and prediction of failures.

Author Contributions: Conceptualization, G.T. and K.K.; methodology, G.T., K.K. and Y.S.; validation, I.K. and Y.S.; formal analysis, G.T.; investigation, K.K. and B.Z.; resources, Y.S.; data curation, B.Z. and Y.S.; writing—original draft preparation, K.K.; writing—review and editing, G.T. and K.K.; visualization, Y.S. and B.Z.; supervision, I.K.; project administration, G.T. and I.K.; funding acquisition, I.K.; editing, consultation and review E.Z. All authors have read and agreed to the published version of the manuscript.

Funding: This study is financed by the European Union—Next Generation EU through the National Recovery and Resilience Plan of the Republic of Bulgaria, project № BG-RRP-2.004-0005, and by project КП-06-Н67/8: Development of a fluid–structural methodology for the study and modernization of turbomachines through virtual prototyping—HydraTech.

Data Availability Statement: The data presented in this study are available on request from the corresponding author. The data are not publicly available due to their relation to public funding specifics.

Conflicts of Interest: The authors declare no conflicts of interest.

References

1. Available online: https://www.nek.bg/index.php/en/ (accessed on 5 January 2025).
2. Available online: https://wwtech.com.pl/en/2024/09/11/chaira-pump-storage-hpp-bulgaria/ (accessed on 1 April 2025).
3. Available online: https://balkangreenenergynews.com/bulgaria-to-replace-broken-turbines-in-chaira-pumped-storage-hydropower-plant/ (accessed on 5 January 2025).
4. Available online: https://en.wikipedia.org/wiki/List_of_hydroelectric_power_station_failures (accessed on 1 April 2025).
5. Yasuda, M.; Watanabe, S. How to Avoid Severe Incidents at Hydropower Plants. *Int. J. Fluid Mach. Syst.* **2017**, *10*, 296–306. [CrossRef]
6. Available online: https://en.wikipedia.org/wiki/Sayano-Shushenskaya_power_station_accident (accessed on 1 April 2025).
7. Nicolet, C.; Avellan, F.; Allenbach, P.; Sapin, A.; Simond, J.-J. New Tool for the Simulation of Transient Phenomena in Francis Turbine Power Plants. In Proceedings of the Hydraulic Machinery and Systems, 21st IAHR Symposium, Lausanne, Switzerland, 9–12 September 2002. Available online: https://www.powervision-eng.ch/Profile/Publications/pdf/IAHR_2002.pdf (accessed on 1 April 2025).
8. Available online: https://www.hydropower.org/sediment-management-case-studies/nepal-kali-gandaki (accessed on 1 April 2025).
9. Aliabadi; Jam, F.; Ghahremani, H. Crack Formation on Francis Runners for Karun III Hydroelectric Project: Causes and Solutions. In Proceedings of the 1st International and 3rd National Conference on Dams and Hydropower, Tehran, Iran, 8–13 February 2012. Available online: https://civilica.com/doc/138473/ (accessed on 1 April 2025).
10. Alvarez, A. Solving a Problem Related to Shear Pin Failure. Hydro Review 2011. Available online: https://www.hydroreview.com/world-regions/latin-america/solving-a-problem-related-to-shear/ (accessed on 1 April 2025).
11. Halipchuk, P.; Penner, K.; Smith, S. Machining Solutions: Repairing Cracked Turbine Shafts at Jenpeg. HydroWorld.com. Available online: https://www.renewableenergyworld.com/baseload/hydropower/machining-solutions-repairing-cracked-turbine-shafts-at-jenpeg/ (accessed on 1 April 2025).
12. Price, J.W.H. The failure of the Dartmouth turbine casing. *Int. J. Press. Vessel. Pip.* **1998**, *75*, 559–566. [CrossRef]
13. Gao, X.; Fu, D.; Wu, H. Embedment of Steel Spiral Casings in Concrete: Lessons from a Structural Deformation Accident in China. *Appl. Sci.* **2022**, *12*, 8395. [CrossRef]
14. Ma, Z.; Zhang, C. Static and dynamic damage analysis of mass concrete in hydropower house of Three Gorges Project. *Trans. Tianjin Univ.* **2010**, *16*, 433–440. [CrossRef]
15. Todorov, G.; Kralov, I.; Kamberov, K.; Sofronov, Y.; Zlatev, B. Failure mechanisms of stay vanes in a PHES Francis turbine unit. In Proceedings of the 13th International Scientific Conference "TechSys 2024"—Engineering, Technologies and Systems, Plovdiv, Bulgaria, 18 May 2024.
16. Todorov, G.; Kralov, I.; Kamberov, K.; Zahariev, E.; Sofronov, Y.; Zlatev, B. An Assessment of the Embedding of Francis Turbines for Pumped Hydraulic Energy Storage. *Water* **2024**, *16*, 2252. [CrossRef]

17. Todorov, G.; Kralov, I.; Kamberov, K.; Sofronov, Y.; Zlatev, B.; Zahariev, E. Investigation and Identification of the Causes of the Unprecedented Accident at the "Chaira" Pumped Hydroelectric Energy Storage. *Water* **2024**, *16*, 3393. [CrossRef]
18. Hanel, B. FKM-Guideline. In *Analytical Strength Assessment of Components in Mechanical Engineering*; VDMA Verlag GmbH: Frankfurt, Germany, 2003.
19. Chryssanthopoulos, M.; Righiniotis, T. Fatigue reliability of welded steel structures. *J. Constr. Steel Res.* **2006**, *62*, 1199–1209. [CrossRef]
20. Liu, X.; Luo, Y.; Wang, Z. A review on fatigue damage mechanism in hydro turbines. *Renew. Sustain. Energy Rev.* **2016**, *54*, 1–14. [CrossRef]
21. Lyutov, A.; Kryukov, A.; Cherny, S.; Chirkov, D.; Salienko, A.; Skorospelov, V.; Turuk, P. Modeling of a Francis Turbine Runner Fatigue Failure Process Caused by Fluid-Structure Interaction. *IOP Conf. Ser. Earth Environ. Sci.* **2016**, *49*, 072012. [CrossRef]
22. Presas, A.; Luo, Y.; Wang, Z.; Guo, B. Fatigue life estimation of Francis turbines based on experimental strain measurements: Review of the actual data and future trends. *Renew. Sustain. Energy Rev.* **2019**, *102*, 96–110. [CrossRef]
23. Biner, D.; Alligné, S.; Nicolet, C.; Dujic, D.; Münch-Alligné, C. Numerical fatigue damage analysis of a variable speed Francis pump-turbine during start-up in generating mode. In *IOP Conference Series: Earth and Environmental Science*; IOP Publishing: Bristol, UK, 2022; Volume 1079, p. 012079. Available online: https://iopscience.iop.org/article/10.1088/1755-1315/1079/1/012079 (accessed on 1 April 2025).
24. Zhang, Q.-L.; Hu, L.; Gao, X.-F. Fatigue life prediction of steel spiral cases in pumped-storage power plants: Factors to be considered. *Eng. Fail. Anal.* **2024**, *157*, 107908. [CrossRef]
25. Souza, R.Q.; Álvares, A.J. FMEA and FTA Analysis for Application of the Reliability Centered Maintenance Methodology: Case Study on Hydraulic Turbines. ABCM Symposium Series in Mechatronics. 2008; Volume 3, pp. 803–812. Available online: https://www.abcm.org.br/symposium-series/SSM_Vol3/Section_VII_Emerging_Technologies_and_Applications/SSM3_VII_10.pdf (accessed on 1 April 2025).
26. Peeters, J.F.W.; Basten, R.J.I.; Tinga, T. Improving failure analysis efficiency by combining FTA and FMEA in a recursive manner. *Reliab. Eng. Syst. Saf.* **2018**, *172*, 36–44. [CrossRef]
27. InfraSpeak Team. FTA vs FMEA: What Are the Differences? 23 October 2020. Available online: https://blog.infraspeak.com/fta-vs-fmea/ (accessed on 1 April 2025).
28. Standard: DIN EN ISO 12100:2011-03, Safety of Machinery—General Principles for Design—Risk Assessment and Risk Reduction, Publication Date 2011-03, Original Language German, Pages 96. Available online: https://www.dinmedia.de/en/standard/din-en-iso-12100/128264334 (accessed on 1 April 2025).
29. Flynn, A.M. Hazard Risk Assessment, Chapter in the Book. In *Introduction to Environmental Management*, 2nd ed.; Theodore, M.K., Theodore, L., Eds.; CRC Press: Boca Raton, FL, USA, 2021; pp. 1–11. Available online: https://www.taylorfrancis.com/chapters/edit/10.1201/9781003171126-44/hazard-risk-assessment-ann-marie-flynn (accessed on 1 April 2025).
30. TheSafetyMaster. *Hazard Identification and Risk Assessment*; Study Report Consultancy and Training Services; TheSafetyMaster: Gurugram, India, 2024. Available online: https://www.thesafetymaster.com/hazard-identification-and-risk-assessment-study-report-consultancy-and-training-services-by-thesafetymaster/ (accessed on 1 April 2025).
31. Standard DIN EN ISO 14121-1:2007-12 Safety of Machinery—Risk Assessment—Part 1: Principles (ISO 14121-1:2007); German Version Sicherheit von Maschinen—Risikobeurteilung—Teil, Pages 35. Available online: https://www.dinmedia.de/en/standard/din-en-iso-14121-1/94405328 (accessed on 1 April 2025).
32. Quality-One International. Introduction to Failure Mode, Effects & Criticality Analysis (FMECA). Available online: https://quality-one.com/fmeca/ (accessed on 1 April 2025).
33. Military Handbook. Reliability Prediction of Electronic Equipment. 1991. Available online: https://www.navsea.navy.mil/Portals/103/Documents/NSWC_Crane/SD-18/Test%20Methods/MILHDBK217.pdf (accessed on 1 April 2025).
34. Rausand, M.; Haugen, S. *Risk Assessment: Theory, Methods, and Applications*, 2nd ed.; Wiley: Hoboken, NJ, USA, 2020; ISBN 978-1-119-37722-1.

energies

MDPI

Article

Multi-Objective Sensitivity Analysis of Hydraulic–Mechanical–Electrical Parameters for Hydropower System Transient Response

Yongjia Li [1,2], Yixuan Guo [1,2], Ming Li [1,2], Liuwei Lei [1,2], Huaming Hu [1,2], Diyi Chen [1,2,*], Ziwen Zhao [1,2,*] and Beibei Xu [1,2,*]

1 Key Laboratory of Agricultural Soil and Water Engineering in Arid and Semiarid Areas, Ministry of Education, Northwest A&F University, Xianyang 712100, China; 18992847498@163.com (Y.L.); wz991230@163.com (Y.G.); samllermultiply@163.com (M.L.); liuweilei@nwafu.edu.cn (L.L.); 17590696850@163.com (H.H.)

2 Institute of Water Resources and Hydropower Research, Northwest A&F University, Xianyang 712100, China

* Correspondence: diyichen@nwafu.edu.cn (D.C.); zhaoziwen@nwsuaf.edu.cn (Z.Z.); xubeibei0413@163.com (B.X.)

Abstract: Hydropower's ability to start up and shut down quickly, combined with its flexible regulation characteristics, effectively alleviates frequency fluctuations caused by new energy sources, ensuring the safe and stable operation of the power system. However, during peak-frequency regulation tasks, the transition processes associated with the startup, shutdown, and load changes introduce frequent shocks to subsystems such as the hydro-turbine, governor, and diversion systems. These shocks pose significant challenges to the safe and stable operation of hydropower plants. Therefore, this study constructs a coupled hydraulic–mechanical–electrical model that incorporates the diversion system, hydro-turbine, governor, generator, and load, based on operational data from a real-world hydropower plant in China. The load increase transition process is selected for parameter sensitivity analysis to evaluate the influence of various structural, operational, and control parameters on unit stability and to identify key parameters affecting stability. The results indicate that the initial load exhibits the highest sensitivity to inversion power peak and rotational speed overshoot, with sensitivity values of 0.14 and 0.0038, respectively. The characteristic water head shows the greatest sensitivity to the inversion power peak time and rotational speed peak time, with values of 0.31 and 0.43, respectively. Additionally, the integration gain significantly influences the rotational speed rise time, with a sensitivity value of 0.30. These findings provide a theoretical basis for optimizing the parameter selection in hydropower plants.

Keywords: hydropower units; transition process; sensitivity; stability

1. Introduction

Hydropower is a crucial component of the future renewable energy mix, characterized by its clean, environmentally friendly nature, flexible scheduling, and low operating costs [1,2]. Developing the hydropower industry is essential for meeting the growing energy demand, optimizing the energy structure, conserving energy, reducing emissions, and ensuring the safety and stability of the power system [3,4]. With its ability to accurately track load changes and dynamically adjust, hydropower is the ideal energy source for regulating grid fluctuations, optimizing capacity allocation, and providing peak load support [5]. However, during frequent peak load and frequency regulation tasks, hydroelectric systems

https://doi.org/10.3390/en18102609

undergo transitional processes, such as startup, shutdown, and load adjustments [6]. In these processes, various system parameters continuously change and may even enter abnormal states. Prolonged operation under such conditions accelerates the degradation of core components, such as the rotor [7], posing significant risks to the safety and reliability of the hydropower station's operation. Furthermore, since a hydropower system is a complex integration of hydraulic, mechanical, and electrical subsystems [8], it is essential to fully account for the interactions and coupling between these subsystems to accurately capture the dynamic response characteristics during transient processes. Therefore, constructing a simulation model that accurately reflects the dynamic response in the transient process and revealing the mechanism of structural design, operating conditions, and speed control parameters on the stability of the unit has become the key to ensuring the safe and stable operation of the hydropower system [9].

Currently, research on the operational stability of hydropower systems can be broadly categorized into two main areas: the development of power system models and the analysis of the influence of operational parameters. In terms of model development, relevant scholars commonly employ numerical simulation methods. Compared to conventional experimental measurements, numerical simulation effectively overcomes limitations in cost, time, and research depth [10]. This approach enables a comprehensive and accurate observation of physical phenomena and parameter variations at critical locations under extreme operating conditions through in-depth system analysis. These advantages are particularly pronounced in materials research, as clearly demonstrated in the studies by EL-SAPA et al. (2022, 2023, 2022) [11–13]. In the context of our research, Celebioglu et al. (2017) [14] employed the commercial software ANSYS Bladegen to investigate cavitation characteristics of hydroelectric units under off-design flow conditions, ultimately achieving cavitation-free operation optimization for an actual turbine runner. Yang et al. (2016) [15] conducted modeling using TOPSYS from a control perspective, investigating the wear mechanisms in hydropower units during primary frequency regulation, and identified effective methods to reduce wear. In mechanical systems, Valentín et al. (2017) [16] developed a hydraulic–mechanical–electrical coupled model using the finite element method to study the relationship between instability and power oscillations under partial load and overload conditions. Zhang et al. (2019) [17] expanded on previous work with a hydraulic–mechanical–electrical–structural model, exploring the vibration behavior of critical components such as guide bearings, rotors, and runners under sudden load increases.

Regarding the influence of parameters on unit operation, relevant scholars have conducted extensive research in three main areas: transient analysis and dynamic studies, optimization and structural analysis of hydraulic systems, and system stability and control strategy optimization. Ma et al. (2024) [18] proposed a framework for transient analysis under parameter uncertainty, encompassing integrated modeling, uncertain transient analysis, and transient calibration. Their findings highlighted the significant impact of six hydraulic generator parameters on transient characteristics, with casing pressure being particularly critical. Zhu et al. (2022) [19] employed trajectory sensitivity analysis to examine the sensitivity of the model's state variables and their primary and secondary relationships to subsystem parameter changes. They demonstrated that state variables are highly sensitive to variations in the hydraulic turbine torque transfer coefficient, unit inertia time constant, and the proportional gain of the controller during guide vane opening. In the area of structural parameter analysis, Lei et al. (2021) [20] utilized the characteristic line method to develop a hydraulic power generation system and explored startup optimization strategies for both symmetric and asymmetric penstock structures. Li et al. (2021) [21] analyzed engineering case studies to compare the operating characteristics of different tailwater pressure chamber configurations under varying conditions. Regarding system stability and control strategy optimization, Liao et al. (2022) [22] investigated the influence mechanisms of four key hydraulic dynamic time constants on nonlinear system stability with backlash. Using the Nyquist stability criterion and numerical simulation, they provided theoretical insights into hydraulic generator dynamic regulation stability. Liu et al. (2017) [23] applied the pole placement method to optimize PID parameters, addressing excessive low-frequency oscillations caused by a low proportional-to-integral gain ratio. Additionally, Singh et al. (2013) [24] utilized genetic algorithms to optimize water turbine speed control parameters during load variation transitions, achieving improved dynamic performance of the overall system.

In summary, although significant progress has been made in the research aimed at improving the stability of hydropower units, two main shortcomings remain:

1. Hydropower system Modeling: Some researchers have failed to adequately consider the influence of mechanical and electrical systems in their models, relying solely on existing commercial software for stability analysis of the hydraulic system. Moreover, while some scholars have established relatively complete hydraulic–mechanical–electrical coupled models, these models tend to overemphasize the coupling relationships between the mechanical and electrical subsystems, neglecting the complexity of the hydraulic system itself and its impact on the overall system stability.
2. Parameter Sensitivity Analysis: Current research often focuses on a single category of parameters (e.g., structural, operating, or control parameters), leading to insufficient depth and breadth in the analysis of parameter impacts. There is a lack of comprehensive cross-comparison of multiple parameters, which hinders the identification of the relative importance of different parameter types in the operation of hydropower units and the quantification of the impact of core parameters on unit stability indicators.

To address the gap, this paper constructs a hydraulic–mechanical–electrical coupled model for a real hydropower system in China, incorporating the intake system, turbine, governor, generator, and load, while fully considering the characteristics and coupling relationships among the hydraulic, mechanical, and electrical subsystems. Then, focusing on structural design parameters, operating conditions, and control parameters, a parameter sensitivity analysis was performed during the load increase transition process to explore the impact of different structural, operational, and control parameters on unit stability. The sensitivity of these parameters is analyzed and ranked, enabling the identification of the core parameters that affect unit stability.

The paper is structured as follows: Section 2 presents the hydropower system model based on a real hydropower system in China, coupling the intake system, turbine, governor, generator, and load. Section 3 includes the stability analysis of unit operation and sensitivity analysis of structural, operating, and control parameters. Sections 4 and 5 discuss and summarize the findings.

2. Hydropower Generator System Modeling

This section employs a real hydropower system in China as the subject of modular modeling. The coupled model consists of several subsystems, including a diversion system model constructed using the method of characteristics (MOC), a generator and load model, a PID speed regulator model for single-machine load frequency control, and a hydro-turbine model derived from the hydro-turbine's full characteristic curve. Based on the system's internal structure and external interaction mechanisms, the aforementioned subsystems are nonlinearly coupled through parameters such as guide vane opening, rotational speed, flow rate, water head, turbine prime mover torque, and generator electromagnetic torque, resulting in a nonlinear coupled model of the hydropower system.

2.1. Diversion System

A hydropower station's diversion system consists of upstream and downstream reservoirs, diversion tunnels, a surge tank, and pressure pipes. The water hammer equation for a pressure pipe is a quasi-linear partial differential equation system involving flow velocity V and water head H. This paper solves the water hammer equation using MOC, and the schematic diagram of the hydropower generation system is shown in Figure 1.

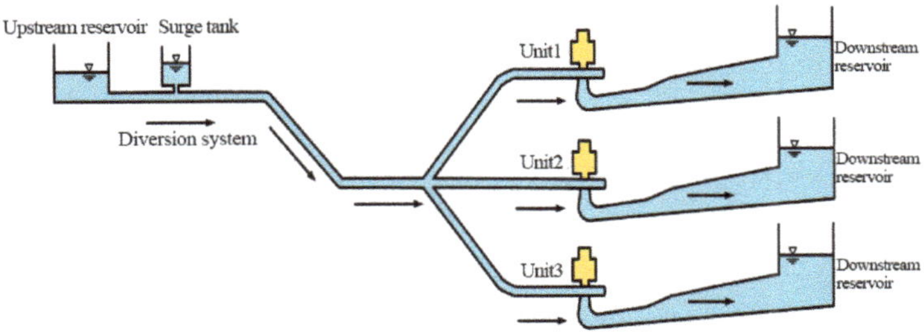

Figure 1. Schematic diagram of the hydropower generation system.

This paper selects a non-constant flow pipe model and applies the MOC to solve the model of the diversion system. Figure 2 presents the schematic diagram of the MOC. In the $x - t$ plane, C^+ and C^- are referred to as characteristic lines, and ΔX represents the equally divided length of the pipe; Δt is the time step. The simplified expression is given by cities [25] (Figure 2).

$$C^+ \Rightarrow \begin{cases} \frac{1}{a}\frac{dH}{dt} + \frac{1}{g}\frac{dV}{dt} + \frac{f|V|V}{2gD} + \frac{1}{a}V\sin\sigma = 0 \\ \frac{dX}{dt} = V + a \end{cases} \\ C^- \Rightarrow \begin{cases} -\frac{1}{a}\frac{dH}{dt} + \frac{1}{g}\frac{dV}{dt} + \frac{f|V|V}{2gD} - \frac{1}{a}V\sin\sigma = 0 \\ \frac{dX}{dt} = V - a \end{cases} \tag{1}$$

where Q_p and H_p are the flow and water head at point p in the pipe at the time $t = t_0 + \Delta t$, respectively; C_a, C_p, and C_n are the intermediate variables; A denotes the pipe's cross-sectional area. Q_d and Q_u represent the flow rates at points d and u in the pipe at time $t = t_0$. H_d and H_u are the water heads at points d and u in the pipe at time $t = t_0$. When the boundary conditions at the ends of each part of the pipeline are known, the individual states of the point p at $t = t_0 + \Delta t$ can be determined by solving Equation (1). The selected time step for the calculation must satisfy the Courant stability condition, i.e., $\Delta t \leq \Delta x/a$.

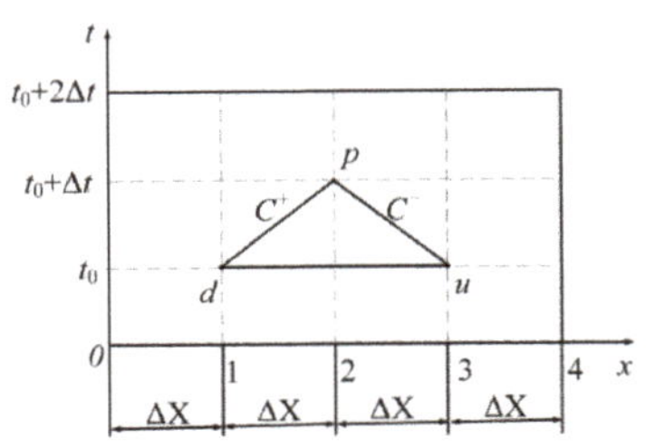

Figure 2. The schematic diagram of the method of characteristics.

When calculating the transition process, only one characteristic line equation is satisfied at the upstream and downstream nodes, as well as the connecting nodes of each pipe section. Therefore, the solution is obtained by combining the boundary conditions. The boundary conditions required in this paper include upstream and downstream reservoirs, surge tank, elbow pipes, bifurcated pipe, and hydro-turbine [26]. Figure 3 illustrates a schematic diagram of the boundary conditions.

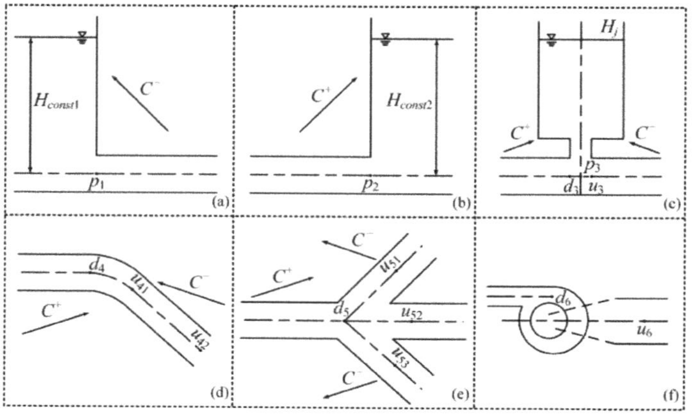

Figure 3. The schematic diagram of boundary conditions for each node of the water diversion system: (**a**) upstream reservoir; (**b**) downstream reservoir; (**c**) surge tank; (**d**) elbow pipe; (**e**) bifurcated pipe; (**f**) hydro-turbine.

1. Upstream and downstream reservoirs

Assuming that the water levels in the upstream and downstream reservoirs remain constant during the transient flow period, and ignoring the water head loss at the pipe inlet, it can be concluded that the water levels in both reservoirs are constant. The water level at the upstream reservoir outlet follows the negative characteristic line equation C^-, while the downstream reservoir inlet follows the positive characteristic line equation C^+. The boundary conditions for the upstream and downstream reservoirs are as follows:

$$\begin{cases} H_{p1} = H_{const1} \\ Q_{p1} = C_{n1} + C_{a1}H_{p1} \\ H_{p2} = H_{const2} \\ Q_{p2} = C_{p2} - C_{a2}H_{p2} \end{cases} \tag{2}$$

where Q_{p1} and Q_{p2} are the flow rates at points p_1 and p_2, respectively; H_{p1} and H_{p2} are the water heads at points p_1 and p_2, respectively; H_{const1} and H_{const2} are the water heads of the upstream and downstream reservoirs, respectively; C_{n1}, C_{p2}, C_{a1}, and C_{a2} are intermediate variables.

2. Surge tank

The primary function of the surge tank is to reduce water hammer pressure in the pipeline. It can be classified into various types, including impedance surge tanks, air cushion surge tanks, differential surge tanks, variable cross-section surge tanks, and others. In this paper, the widely used impedance surge tank is selected, as shown in Figure 3c, and the C^+ and C^- equations can be derived for the d_3 and u_3 sections, respectively.

$$\begin{cases} Q_{d3} = C_{p3} - C_{a31}H_{d3} \\ Q_{u3} = C_{n3} + C_{a32}H_{u3} \end{cases} \tag{3}$$

From the continuity equation and the energy equation, assuming no water head loss, it can be obtained:

$$\begin{cases} Q_{d3} = Q_j + Q_{u3} \\ H_{d3} = H_{u3} = H_{p3} \end{cases} \tag{4}$$

The water level equation at point p_3 is expressed as follows:

$$H_{p3} = H_j + R_j \frac{Q_j |Q_j|}{A_{wj}^2} \tag{5}$$

The relationship between the water level and flow rate in the surge tank is as follows:

$$H_j = H_{j,\Delta t} + \frac{(Q_j + Q_{j,\Delta t})\Delta t}{A_{wj}^2} \tag{6}$$

where Q_{d3} and Q_{u3} are the flows at points d_3 and u_3, respectively. H_{d3} and H_{u3} are the water heads at the points d_3 and u_3, respectively; C_{p3}, C_{n3}, C_{a31}, and C_{a32} are intermediate variables. Q_j and Q_j, Δt represent the current and previous flow rates entering the surge tank. H_j and H_j, Δt represent the surge tank's current and previous water heads. Hp_3 is the water head at point p_3; R_j is the hydraulic loss coefficient of the surge tank's impedance hole; A_{wj} is the impedance hole area; and A_j is the surge tank's cross-sectional area. The impedance surge tank's boundary conditions can be obtained by combining the four equations listed above.

3. Elbow pipe

As shown in Figure 3d, the front end (d_4) and the back end (u_{42}) of the elbow pipe satisfy the C^+ and C^- equations, respectively, and their boundary conditions are:

$$\begin{cases} Q_{d4} = C_{p4} - C_{a41} H_{d4} \\ Q_{u41} = C_{n4} + C_{a42} H_{u41} \\ Q_{u41} = Q_{u42} \\ H_{u41} = H_{u42} + \xi \frac{V_{u42}^2}{2g} \end{cases} \tag{7}$$

where Q_{d4}, Q_{u41}, and Q_{u42} represent the flow rates at points d_4, u_{41}, and u_{42}, respectively; H_{d4}, Hu_{41}, and H_{u42} represent the water heads at points d_4, u_{41}, and u_{42}, respectively; C_{p4}, C_{n4}, C_{a41}, and C_{a42} are intermediate variables, and ξ is the local water head loss coefficient.

4. Bifurcated pipe

As shown in Figure 3e, section d_5 satisfies the C^+ equation and sections u_{51}, u_{52}, and u_{53} satisfy the C^- equation whose boundary conditions are:

$$\begin{cases} Q_{d5} = C_{p5} - C_{a5} H_{d5} \\ Q_{u51} = C_{n51} + C_{a51} H_{u51} \\ Q_{u52} = C_{n52} + C_{a52} H_{u52} \\ Q_{u53} = C_{n53} + C_{a53} H_{u53} \\ H_{d5} = H_{u51} = H_{u52} = H_{u53} = H_p \\ Q_{d5} = Q_{u41} + Q_{u52} + Q_{u53} \end{cases} \tag{8}$$

where Q_{d5}, Q_{u51}, Q_{u52}, and Q_{u53} are the flow rates at points d_5, u_{51}, u_{52}, and u_{53}, respectively; H_{d5}, H_{u51}, H_{u52}, and H_{u53} are the water heads at points d_5, u_{51}, u_{52}, and u_{53}, respectively; C_{p5}, C_{n51}, C_{n52}, C_{n53}, C_{a5}, C_{a51}, C_{a52}, and C_{a53} are intermediate variables; H_p is the water head at point p.

5. Unit

The water head of the hydro-turbine is the difference between the inlet of the volute casing and the outlet of the tailpipe. Without considering the flow rate loss of the hydro-turbine, the boundary condition of the hydro-turbine can be expressed as:

$$\begin{cases} Q_{d6} = Q_{u6} = Q_t \\ Q_{d6} = C_{p6} - C_{a61} H_{d6} \\ Q_{u6} = C_{n6} + C_{a62} H_{u6} \\ H_t = H_{d6} - H_{u6} \end{cases} \tag{9}$$

where Q_{d6} and Q_{u6} are the flow rates at points d_6 and u_6, respectively; H_{d6} and H_{u6} are the water heads at points d_6 and u_6, respectively; C_{p6}, C_{n6}, C_{a61}, and C_{a62} are intermediate variables; and Q_t is the flow rate of the hydro-turbine.

2.2. Hydro-Turbine

The dynamic characteristics of the hydro-turbine need to be transformed into the boundary conditions of the pipeline. The characteristic line equations of the pipe are solved simultaneously with known guide vane openings, rotational speed, and water head. The hydro-turbine characteristic data are converted into characteristic curves through polynomial interpolation, as shown in Figure 4. The formulas for flow rate, torque, and rotational speed are:

$$\begin{cases} Q_{11} = f(n_{11}, a), Q_{u6} = Q_{11} D_1^2 \sqrt{H_t} \\ M_{11} = f(n_{11}, a), M_t = M_{11} D_1^3 H_t \\ n_t = \frac{n_{11}\sqrt{H_t}}{D_1} \end{cases} \tag{10}$$

where Q_{11} is the unit flow rate, M_{11} is the unit torque, n_{11} is the unit rotational speed, and a is the guide vane opening.

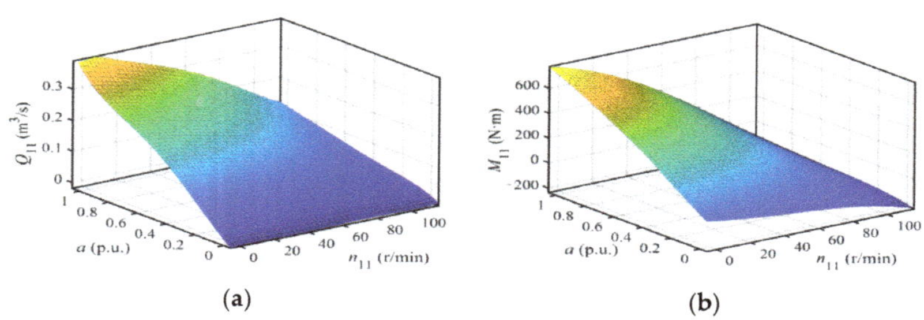

Figure 4. The characteristic curves of hydro-turbine. (**a**) Flow characteristic curve; (**b**) torque characteristic curve.

2.3. Generator and Load

A first-order generator model is chosen. The unit system remains connected to the grid during the load increase transition process, and the generator's electromagnetic torque is not zero [27]. Therefore, the generator and load are modeled as:

$$\begin{cases} M_t - M_g = J \cdot \frac{d\omega}{dt} \\ m_g = C_g + A_g + e_g(n - n_r) \\ n_t = n_{t-\Delta t} + \frac{\Delta t}{T_a}\left[1.5\left(m_{t-\Delta t} - m_{g(t-\Delta t)}\right) - 0.5\left(m_{t-2\Delta t} - m_{g(t-2\Delta t)}\right)\right] \end{cases} \tag{11}$$

where M_t is the hydro-turbine's driving torque; M_g is the generator's electromagnetic torque; ω is the angular velocity; J is the moment of inertia; m_g is the relative electromagnetic torque; C_g is the relative load torque; A_g is the step value of the load torque; n_r is the relative rated rotational speed; e_g is the load self-regulation coefficient; n is the relative rotational speed; T_a is the unit inertia time constant.

2.4. PID Governor

When the unit operates in single-machine load mode, it transitions from load control to frequency control [28]. The input–output response block diagram of the adopted PID governor is shown in Figure 5.

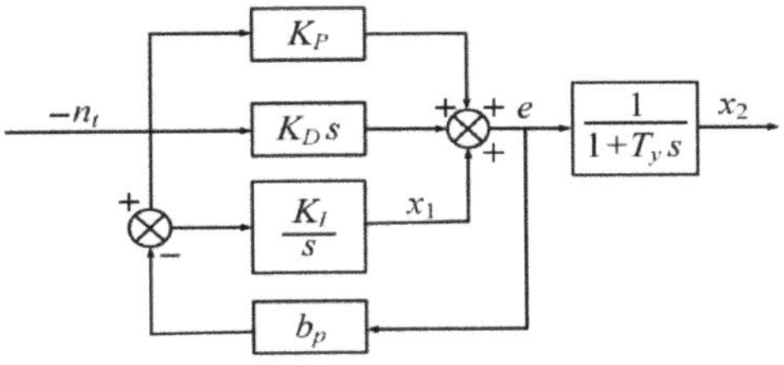

Figure 5. PID governor input–output response block diagram.

The equation of state for the system is:

$$\begin{bmatrix} \dot{x}_1 \\ \dot{x}_2 \end{bmatrix} = \begin{bmatrix} -b_p & 0 \\ 1/T_y & -1/T_y \end{bmatrix} \begin{bmatrix} x_1 \\ x_2 \end{bmatrix} + \begin{bmatrix} b_p K_p - K_I & b_p K_D \\ -K_P/T_y & -K_D/T_y \end{bmatrix} \begin{bmatrix} n_t \\ \dot{n}_t \end{bmatrix} \tag{12}$$

where K_P, K_I, and K_D is the proportional, integral, and differential coefficients of the PID speed controller, respectively; x_1 and e are intermediate variables, and x_2 is the output variable; n_t is the unit rotational speed, T_y is the governor response time, and b_p is the steady-state slip coefficient.

3. Results and Analysis

3.1. Influence of Pipe Structural Parameters on Unit Operation Stability

3.1.1. Main Branch Pipe Diameter Ratio

The selection of the pressure diversion pipe diameter is closely related to the investment cost of the hydropower station. A larger diameter pipe requires a higher initial investment; however, if the diameter is too small, it increases the water hammer pressure at the end of the volute, negatively affecting the stable operation of the unit. During large fluctuations in the transition process, the sudden change in water hammer pressure has a greater impact on the unit. Therefore, selecting a reasonable pressure piping main branch pipe diameter ratio is crucial for both the piping design and the safe, reliable operation of the hydropower station. In this paper, the unit load is increased from 50% to the rated load, with the main branch pipe diameter ratio set at 1.3:1, 1.5:1, 1.7:1, 1.9:1, and 2.1:1, respectively. The effect of different pipe diameter ratios on the system stability of the unit during the transition process is investigated. The dynamic response of the unit under these different pipe diameter ratios is shown in Figure 6.

As shown in Figure 6a, the governor automatically tracks the load change and adjusts the active guide vane to ensure that the unit's rotational speed returns to the vicinity of the rated rotational speed during the load-increasing transition process. Overall, the variation amplitude of the guide vane opening shows an inverse relationship with the main branch pipe diameter ratio. As the diameter ratio decreases, the curve rising rate, peak value, and overshoot all increase. For instance, under diameter ratios from 1.7:1 to 2.1:1, the guide vane openings stabilize around 0.95 p.u., while under the 1.3:1 diameter ratio condition, it reaches 1.03 p.u., maintaining approximately 0.02 p.u. higher after stabilization. In Figure 6b, as the main branch pipe diameter ratio of the hydropower generation system decreases, the unit water head decreases further, and the time required to reach stabilization increases. When the diameter ratio decreases to 1.3:1, the maximum drop in unit water head occurs, with a decrease of 14.98% compared to the rated water head. The variation patterns of unit rotational speed and power with different main branch diameter ratios are shown in Figure 6c,d, respectively. Notably, the trend of unit rotational speed (Figure 6c) closely resembles that of the unit water head (Figure 6b), while the unit power (Figure 6d) follows a similar pattern to the guide vane opening (Figure 6a). Table 1 presents the regulating performance indicators for different main diameter ratios of the main penstocks. The table shows that when the diameter ratio is 1.3:1, the unit experiences the greatest rotational speed and power fluctuations, with a rotational speed regulation time of 34.78 s, a rotational speed overshoot of 0.180, a rotational speed rise time of 14.3 s, a rotational speed peak time of 26.67 s, an inversion power peak of 6.00 MW, and an inversion power peak time of 19.51 s. Notably, the rotational speed regulation time is most significantly affected by changes in the main branch pipe diameter ratio. The maximum rotational speed regulation time (at 1.3:1 diameter ratio) differs from the minimum (at 1.9:1 diameter ratio) by 13.86 s, representing a year-on-year increase of 66.25%. As a result, the smaller the main branch pipe diameter ratio of the diversion pipeline during the unit's load-increase transition process, the greater the fluctuations in water head, rotational speed, and power, leading to worse stability in the transition process.

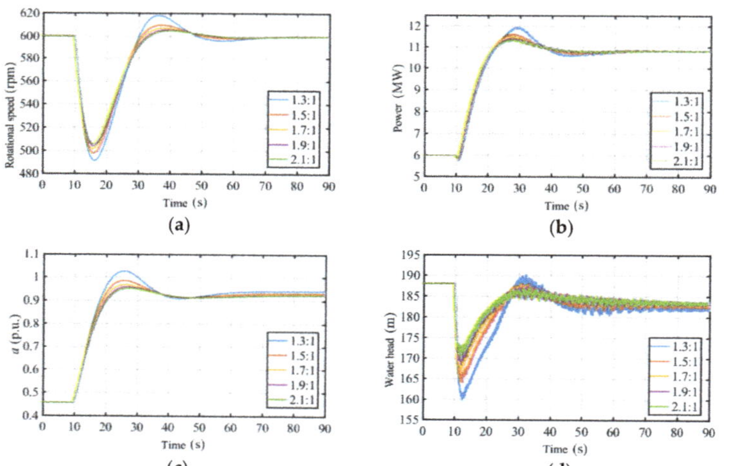

Figure 6. Dynamic response of the unit under different pipe diameter ratios. (**a**) Guide vane opening; (**b**) unit water head; (**c**) unit rotational speed; (**d**) unit power.

Table 1. Regulation performance index under different main branch pipe diameter ratios.

Pipe Diameter Ratio	Rotational Speed Regultion Time	Rotational Speed Overshoot	Rotational Speed Rise Time	Rotational Speed Peak Time	Inversion Power Peak	Inversion Power Peak Time
1.3:1	34.78	0.180	14.03	26.67	6.00	19.51
1.5:1	33.56	0.170	15.02	27.60	5.67	18.47
1.7:1	31.77	0.163	16.09	28.57	5.53	18.37
1.9:1	20.92	0.159	17.02	29.93	5.46	17.44
2.1:1	21.35	0.156	17.33	30.16	5.41	18.26

3.1.2. Surge Tank Location

The location of the surge tank takes into account factors such as the role of the unit in the power system, topography, layout of the diversion system, and economic considerations. Therefore, it is necessary to analyze the results of unit system regulation, protection calculations, and operating conditions during the design process, while synthesizing the influence of various factors to select the optimal location for the surge tank. The pipeline length ratios before and after the surge tank are 1.5:1, 2:1, 2.5:1, 3:1, and 3.5:1, respectively, to investigate the effect of different surge tank positions on the stability of the unit during the load-increase transition process. The guide vane opening, unit water head, unit rotational speed, and unit power at different surge tank positions are shown in Figure 7.

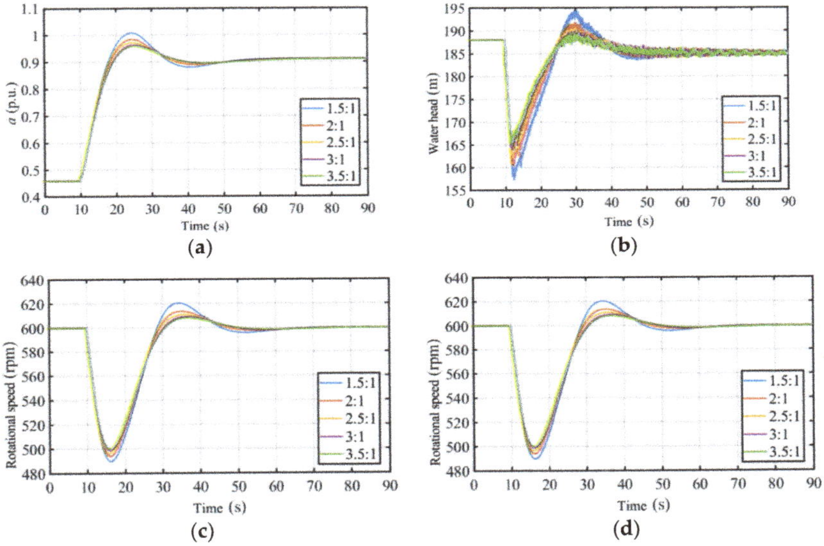

Figure 7. Dynamic response of the unit under different surge tank locations. (**a**) Guide vane opening; (**b**) unit water head; (**c**) unit rotational speed; (**d**) unit power.

As shown in Figure 7a,b, when the load command is issued, the movable guide vanes open quickly, and the unit water head decreases sharply. When the pipeline length ratio before and after the surge tank is 1.5:1, the change in the guide vane opening is the most dramatic, as is the decrease in water head, which is 16.31% lower than the rated water head. Combined with Figure 7c,d and Table 2, it can be seen that the maximum fluctuations in unit rotational speed and power occur when the pipeline length ratio before and after the surge tank is 1.5:1, with a rotational speed regulation time of 32.54 s, a rotational speed overshoot of 0.184, a rotational speed rise time of 12.64 s, a rotational speed peak time of 24.92 s, an inversion power peak of 6.25 MW, and an inversion power peak time of 18.31 s. Moreover, the rotational speed peak time exhibits the most pronounced sensitivity to variations in the pipeline length ratio. The maximum rotational speed peak time (at 3.5:1 pipeline length ratio) differs from the minimum (at 1.5:1 pipeline length ratio) by 12.5 s, corresponding to a significant year-on-year increase of 50.16%. Therefore, the closer the surge tank is to the upstream reservoir, the greater the fluctuations in the unit's water head, rotational speed, and power, and the worse the unit's stability.

Table 2. Regulation performance index under different surge tank locations.

Location of the Surge Tank	Rotational Speed Regulation Time	Rotational Speed Overshoot	Rotational Speed Rise Time	Rotational Speed Peak Time	Inversion Power Peak	Inversion Power Peak Time
1.5:1	32.54	0.184	12.64	24.92	6.25	18.31
2.0:1	32.61	0.177	13.35	25.48	5.94	17.58
2.5:1	32.40	0.173	14.03	36.18	5.80	16.59
3.0:1	32.20	0.169	14.57	36.80	5.71	16.79
3.5:1	31.70	0.167	15.00	37.42	5.65	17.34

3.2. The Impact of Operating Conditions on the Stability of the Unit Operation

3.2.1. Initial Load

Hydropower units operate less efficiently when deviating from their rated conditions to accommodate the regulation of intermittent energy fluctuations. The degree of instability is different for different operating conditions. Therefore, 50% rated load, 60% rated load, 70% rated load, 80% rated load, and 90% rated load are taken to investigate the effect of different initial loads on the system stability of the unit during the load increase transition. Figure 8 displays the unit's dynamic response.

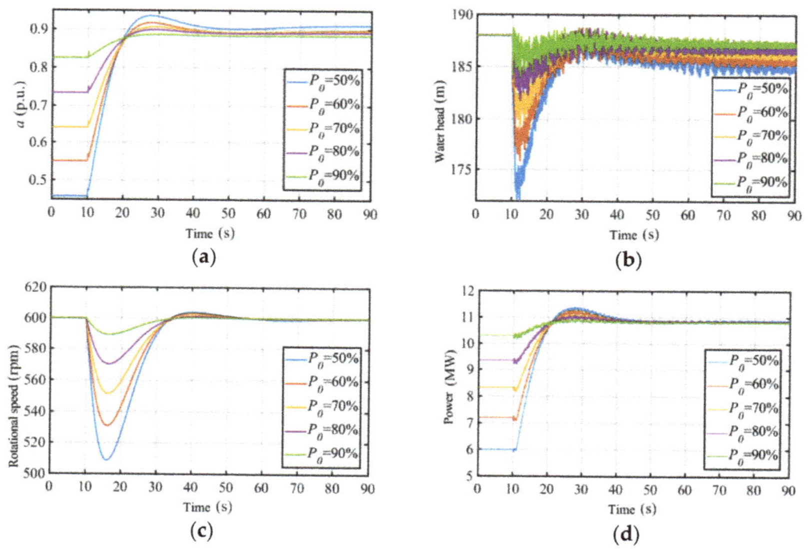

Figure 8. Dynamic response of the unit under different initial loads. (**a**) Guide vane opening; (**b**) unit water head; (**c**) unit rotational speed; (**d**) unit power.

As shown in Figure 8a, the governor opens the movable guide vane to the target opening based on different load changes. The greater the load difference, the steeper the curve, and the larger the rotational speed overshoot. The unit water head drops more and takes longer to stabilize if the initial load is smaller, as shown in Figure 8b. The largest water head drop occurs at 50% load, when the water head drops 8.5% below the rated water head. When combined with Table 3 and Figure 8c,d, it is evident that the unit's power and rotational speed fluctuate most when the initial load is 50% of the rated load. This is especially true when the load is suddenly increased. At this point, the rotational speed regulation time is 21.35 s, the rotational speed overshoot is 0.152 s, the rotational speed rise time is 18.17 s, the rotational speed peak time is 30.04 s, the inversion power peak is 5.38 MW, and the inversion power peak time is 17.84 s. Notably, both rotational speed overshoot and inversion power peak show the highest sensitivity to initial load levels. The maximum values of these parameters occur at 50% initial load, while the minimum values appear at 90% load, with approximately an 11-fold difference between the maximum and minimum values. Overall, the lower the initial load during the load increase, the greater the fluctuations in the unit water head, rotational speed, and power, and the less stable the transition process.

Table 3. Regulation performance index under different initial loads.

Initial Load	Rotational Speed Regulation Time	Rotational Speed Overshoot	Rotational Speed Rise Time	Rotational Speed Peak Time	Inversion Power Peak	Inversion Power Peak Time
50%	21.35	0.152	18.17	30.04	5.38	17.84
60%	20.88	0.115	17.97	30.94	4.06	17.45
70%	20.53	0.081	19.05	31.84	2.77	17.38
80%	19.07	0.050	19.20	31.56	1.71	17.79
90%	14.06	0.018	19.79	30.71	0.63	14.64

3.2.2. Characteristic Water Head

Variations in upstream water inflow and the impact of low water periods cause the unit water head to fluctuate to varying degrees. When the unit system operates at different characteristic water heads, the opening of the movable guide vane varies. In this paper, the unit's characteristic water heads are taken as the rated water head of 188 m, the maximum water head of 214.52 m, and three intermediate water heads of 196.84 m, 205.68 m, and 223.36 m to investigate the impact of different characteristic water heads on the stability of the system. Figure 9 shows the unit's dynamic response under different water heads.

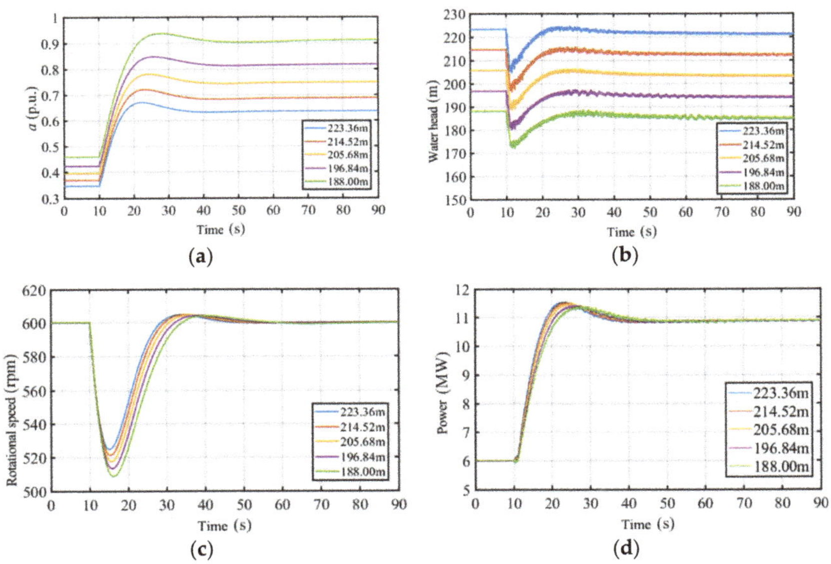

Figure 9. Dynamic response of the unit under different water heads. (**a**) Guide vane opening; (**b**) unit water head; (**c**) unit rotational speed; (**d**) unit power.

As shown in Figure 9a, the smaller the unit water head, the greater the opening of the guide vane at the rated load, resulting in the guide vane moving for a longer duration.

Table 4 lists the regulatory performance indicators for various characteristic water heads. According to Figure 9c,d and Table 4, when the unit water head is 188 m, the unit rotational speed and power fluctuate the most. At this point, the rotational speed regulation time is 16.45 s, the rotational speed overshoot is 0.125, the rotational speed rise time is 13.54 s, the rotational speed peak time is 23.22 s, the inversion power peak is 5.55 MW, and the inversion power peak time is 12.79 s. Notably, all regulatory performance metrics exhibit relatively small variation gradients under different characteristic water head conditions. No significant differences comparable to those induced by initial load parameters, pipeline configurations, or control parameters are observed. Specifically, the deviations between the maximum and minimum values of these metrics remain limited, with neither extreme oscillations nor statistically discernible anomalies manifesting in the dataset. As a result, the smaller the characteristic water head during the unit's load increase, the greater the fluctuations in the unit rotational speed and power, and the less stable the unit is at low water heads.

Table 4. Regulation performance index under different water head.

Characteristic Water Head	Rotational Speed Regulation Time	Rotational Speed Overshoot	Rotational Speed Rise Time	Rotational Speed Peak Time	Inversion Power Peak	Inversion Power Peak Time
223.36	16.45	0.125	13.54	23.22	5.55	12.79
214.52	17.30	0.131	14.25	24.64	5.53	13.41
205.68	18.39	0.138	15.37	26.63	5.47	15.02
196.84	19.63	0.144	16.49	28.67	5.42	16.52
188.00	21.35	0.152	18.17	30.04	5.38	17.84

3.3. *Influence of Control Parameters on the Stability of Unit Operation*

3.3.1. Proportional Gain

When the unit operates in single-unit load-bearing mode, the governing system tracks load changes using frequency control to ensure that the frequency remains close to the rated frequency. The function of the proportional gain is to reflect frequency deviation proportionally; when the load increases, proportional regulation activates to accelerate the regulation and reduce the error. The selection of parameters impacts the system's stability. As a result, the proportional gain is set to 0.6, 0.8, 1.0, 1.2, and 1.4 to investigate the effect of different proportional gains on the unit's stability as the load increases. The dynamic response is depicted in Figure 10.

Figure 10a shows that when the proportional gain is 1.4, the guide vane opens relatively quickly, reducing the time required to reach stability. As shown in Figure 10b, as the proportional gain decreases, the greater the unit water head drop, and the longer it takes to reach stabilization. When the proportional gain is 1.4, the maximum drop in unit water head occurs, with a decrease of 10.21% compared to the rated water head. Figure 10c,d depicts and analyzes the performance indicators of primary frequency modulation, as shown in Table 5. When the proportional gain is 0.6, the unit rotational speed and power fluctuations peak, with a rotational speed regulation time of 38.19 s, a rotational speed overshoot of 0.170, a rotational speed rise time of 14.81 s, a rotational speed peak time of 28.37 s, an inversion power peak of 5.72 MW, and an inversion power peak time of 20.25 s. Notably, the rotational speed regulation time demonstrates the highest sensitivity to variations in proportional gain. The maximum rotational speed regulation time (at proportional gain 0.6) differs from the minimum (at proportional gain 1.4) by 16.84 s, corresponding to a staggering year-on-year increase of 143.69%. Therefore, the smaller the proportional gain, the greater the fluctuations in the unit water head, rotational speed, and power, and the lower the stability during the transition process.

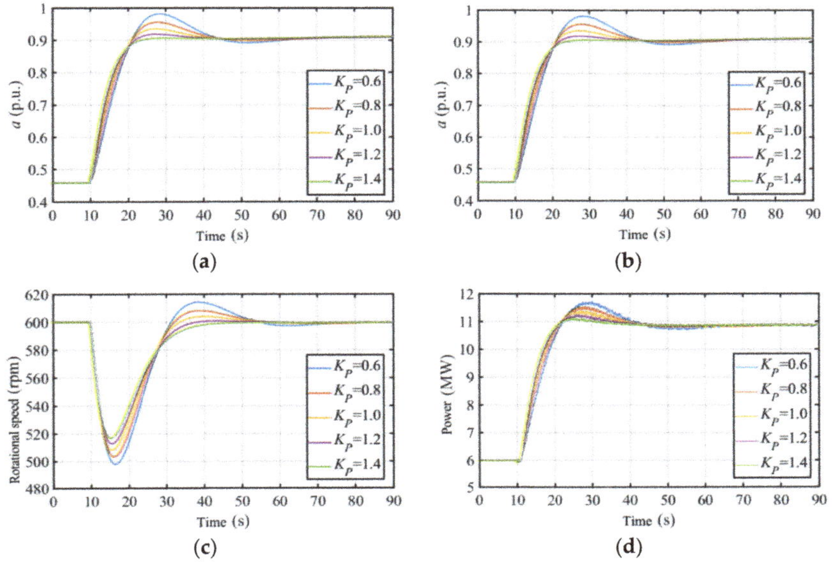

Figure 10. Dynamic response of the unit under different proportional gains. (**a**) Guide vane opening; (**b**) unit water head; (**c**) unit rotational speed; (**d**) unit power.

Table 5. Regulation performance index under different proportional gains.

Proportional Gain	Rotational Speed Regulation Time	Rotational Speed Overshoot	Rotational Speed Rise Time	Rotational Speed Peak Time	Inversion Power Peak	Inversion Power Peak Time
0.6	38.19	0.170	14.81	28.37	5.72	20.25
0.8	34.51	0.160	16.24	28.85	5.53	18.21
1.0	21.35	0.152	18.17	30.04	5.38	17.84
1.2	22.51	0.145	22.46	33.78	5.23	16.69
1.4	24.50	0.138	36.09	36.09	5.12	15.08

3.3.2. Integral Gain

Compared to proportional gain, integral gain maintains the unit's frequency near the rated frequency and helps eliminate residual differences. The integral gain of the unit's governing system is set to 0.6, 0.8, 1.0, 1.2, and 1.4 to investigate the impact of different integral gains on the stability of the unit. The dynamic response is shown in Figure 11.

Figure 11a shows that when the integral gain is 0.29, the guide vane responds quickly, but the overshoot is large, and the time to reach stability is longer. Figure 11b shows that as the integral gain increases, the drop in the unit water head becomes greater, and the time required to reach stability increases. The maximum drop in the unit water head occurs at 0.29, with an 8.83% decrease compared to the rated water head. Figure 11c,d and Table 6 show that when the integral gain is 0.29, the fluctuation in unit rotational speed and power is maximal, with a rotational speed regulation time of 18.87 s, a rotational speed overshoot of 0.149, a rotational speed rise time of 14.77 s, a rotational speed peak time of 26.65 s, an inversion power peak of 5.57 MW, and an inversion power peak time of 16.36 s. Notably, the most sensitive performance parameter to integral gain variations aligns with the effect of proportional gain—the rotational speed rise time. However, contrary to previous observations, the maximum rotational speed rise time (26.27 s) occurs at the minimum integral gain (0.21), whereas the minimum value (14.77 s) appears at the maximum integral gain (0.29), resulting in an absolute difference of 11.5 s and a year-on-year increase of 77.86%. To summarize, as the load on the unit system increases, the higher the integral gain, the greater the variation in the unit water head, rotational speed, and power, and the worse the unit's stability.

Table 6. Regulation performance index under different integral gains.

Integral Gain	Rotational Speed Regulation Time	Rotational Speed Overshoot	Rotational Speed Rise Time	Rotational Speed Peak Time	Inversion Power Peak	Inversion Power Peak Time
0.21	25.45	0.155	26.27	37.79	5.19	19.26
0.23	23.17	0.153	21.31	33.00	5.25	17.41
0.25	21.35	0.152	18.17	30.04	5.38	17.84
0.27	20.01	0.151	16.28	27.67	5.46	17.14
0.29	18.87	0.149	14.77	26.65	5.57	16.36

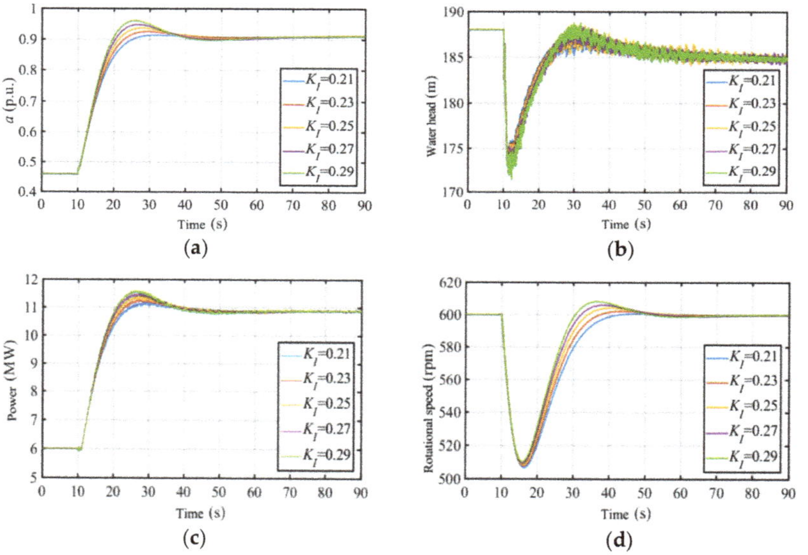

Figure 11. Dynamic response of the unit under different integral gains. (**a**) Guide vane opening; (**b**) unit water head; (**c**) unit rotational speed; (**d**) unit power.

3.4. Sensitivity Analysis of Unit Dynamic Response

The previous section examined the impact of structural design parameters, operating conditions, and system control parameters on the unit's operating characteristics during load increases. Based on the results, the initial input parameters were selected within a ±10% range of values. The effect of each 1% variation in input parameters on operating characteristics was measured. The sensitivity of each parameter to various performance indicators was examined, and the influence of different structural-operating-control input parameters on the unit system's operating characteristics was compared. This analysis serves as a reference for structural design and the safe and stable operation of hydropower systems. Figures 12–17 depict sensitivity heat maps and radar charts for the rotational speed regulation time, rotational speed overshoot, rotational speed rise time, rotational speed peak time, inversion power peak, and inversion power peak time for various structural operating control input parameters.

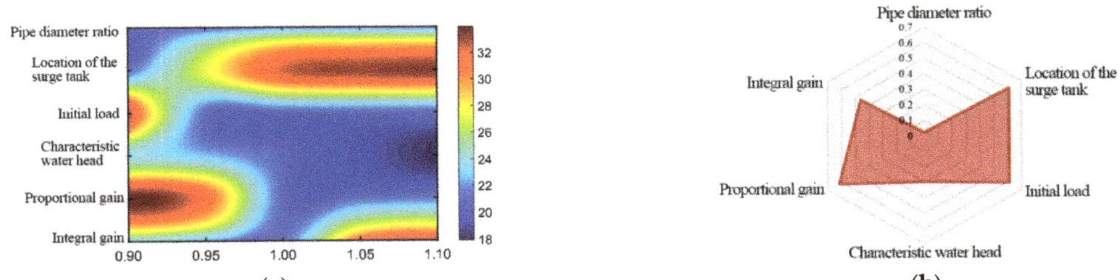

Figure 12. Sensitivity of different input parameters to the rotational speed regulation time. (**a**) Sensitivity heat map; (**b**) radar map.

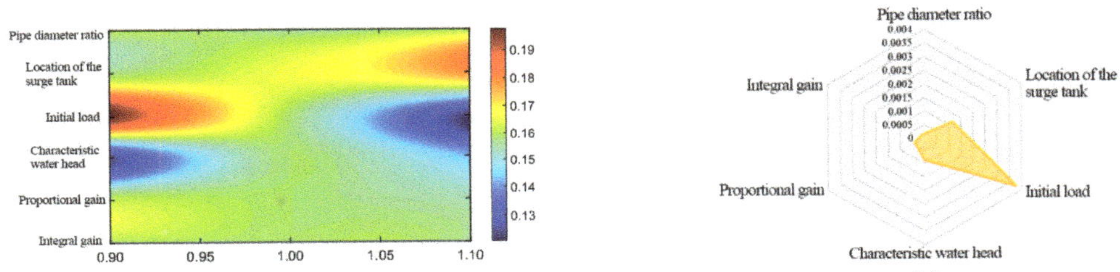

Figure 13. Sensitivity of different input parameters to the rotational speed overshoot. (**a**) Sensitivity heat map; (**b**) radar map.

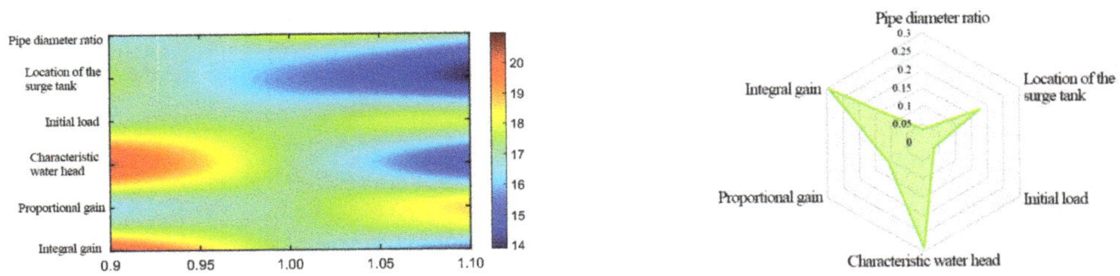

Figure 14. Sensitivity of different input parameters to the rotational speed rise time. (**a**) Sensitivity heat map; (**b**) radar map.

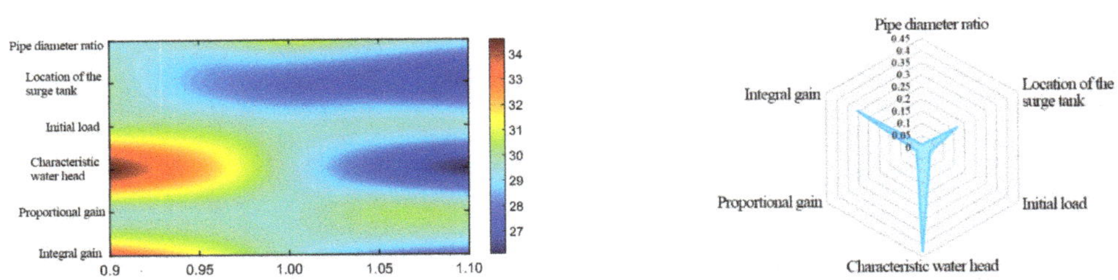

Figure 15. Sensitivity of different input parameters to the rotational speed peak time. (**a**) Sensitivity heat map; (**b**) radar map.

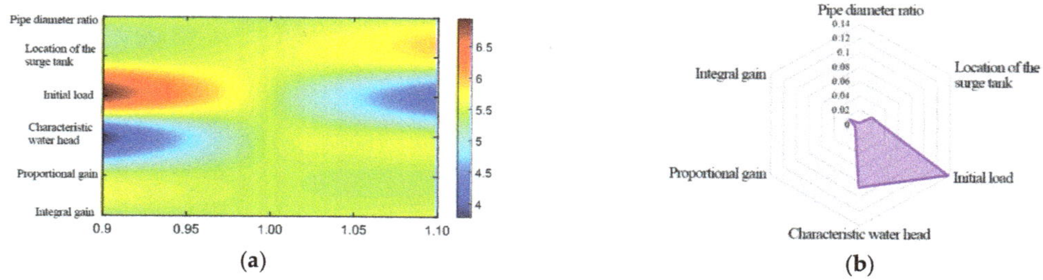

Figure 16. Sensitivity of different input parameters to the inversion power peak. (**a**) Sensitivity heat map; (**b**) radar map.

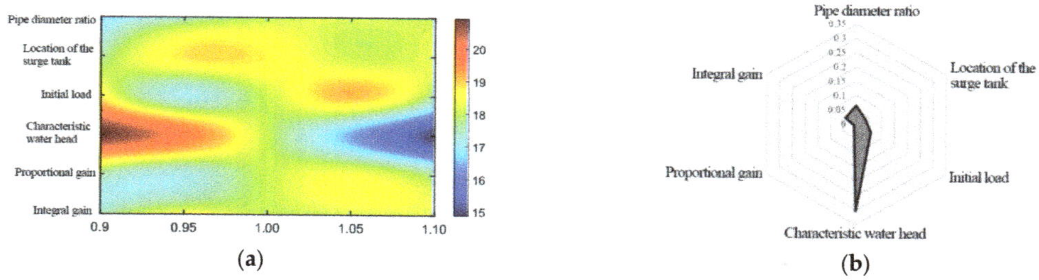

Figure 17. Sensitivity of different input parameters to the inversion power peak time. (**a**) Sensitivity heat map; (**b**) radar map.

The sensitivity of the unit's regulation performance indicators to various structural-operational-control characteristics varies significantly, as shown in Figures 12–17. Among them, the proportional gain is the most sensitive to the rotational speed regulation time, while the pipe diameter ratio is the least sensitive. The sensitivity ranking of each structural-operational-control parameter to the radar chart's rotational speed regulation time is as follows: proportional gain > surge tank location > initial load > integral gain > characteristic water head > pipe diameter ratio. In terms of rotational speed overshoot, the initial load has the highest sensitivity, while the other parameters have relatively low sensitivity, with the following ranking: initial load > surge tank location > pipe diameter ratio > integral gain > characteristic water head > proportional gain. The integral gain has the greatest impact on the rotational speed rise time, followed by the characteristic water head, in this order: integral gain > characteristic water head > surge tank location > proportional gain > pipe diameter ratio > initial load. The characteristic water head has the greatest influence on rotational speed peak time, followed by integral gain, surge tank location, proportional gain, pipe diameter ratio, and initial load. The initial load is the most sensitive to the inversion power peak, with the following sensitivity ranking: initial load > surge tank location > integral gain > characteristic water head > proportional gain > pipe diameter ratio. The characteristic water head is the most sensitive to the inversion power peak time, with the following order: characteristic water head > pipe diameter ratio > initial load > integral gain > surge tank location > proportional gain.

4. Discussion

This study delved into the influence of structural, operational, and control parameters on the stability of hydropower units during the load-increasing transition process by constructing a coupled hydraulic–mechanical–electrical model and conducting parameter sensitivity analysis. The research findings not only lay a theoretical foundation for the optimal selection of hydropower unit parameters but also reveal numerous issues worthy of in-depth exploration and future research directions.

On the one hand, with the large-scale integration of renewable energy, the power system is evolving towards a multi-energy complementary direction. Future research can couple hydropower units with other energy forms, such as wind, solar, and energy storage, to explore their interaction and coordinated operation mechanisms. For example, stochastic differential equations can be introduced to describe the power fluctuations of intermittent renewable energy, and a generalized sensitivity index system for the multi-energy coupling of water–wind–solar–storage can be established to improve the stability theory of the new-type power system dominated by renewable energy.

On the other hand, this study may have overlooked uncertain factors such as equipment aging and environmental factor changes during the modeling process. In the future, uncertainty analysis methods can be introduced to consider the impact of these factors on unit stability, thereby enhancing the robustness and reliability of the model. For instance, methods like Monte Carlo simulation can be used to quantitatively analyze the uncertainty of parameters, providing a more reliable decision-making basis for the design and operation of hydropower stations.

5. Conclusions

In the design and operation of hydropower plants, unsuitable parameters can affect the plant's stability. This paper uses a hydropower station model to investigate the impact of various structural design characteristics, operating condition parameters, and control parameters on the stability of the system during load increases. The main conclusions are as follows:

1. Increasing the main branch pipe diameter ratio and the distance between the surge tank and the upstream reservoir improves stability during the transition. Among these factors, the main branch pipe diameter ratio is most sensitive to the inversion power peak time, while the surge tank's position shows strong sensitivity to the rotational speed regulation time.
2. A larger initial load and characteristic water head enhance the stability of the hydropower plant during the load increase transition process. Among these, the initial load shows strong sensitivity to rotational speed overshoot and inversion power peak, while the characteristic water head is highly sensitive to the rotational speed rise time, rotational speed peak time, and inversion power peak time.
3. Lowering the proportional gain and increasing the integral gain reduces the stability of the hydropower plant system during the transition process. The sensitivity analysis shows that the proportional gain (K_p) is highly sensitive to the rotational speed regulation time, while the integral gain (K_i) strongly affects the rotational speed rise time.

Author Contributions: Y.L. proposed the research framework and technical roadmap, conducted in-depth analysis with data investigation, optimized the methodology, and wrote the original manuscript draft. Y.G. and M.L. contributed to data processing, indicator calculations, and visualization, with Y.G. drafting Figures 6–11 and M.L. drafting Figures 12–17. L.L. drafted Figures 1 and 4, and all tables (Tables 1–6), while H.H. designed Figures 2, 3 and 5. Y.G. and H.H. further coordinated the manuscript revision and formatting adjustments. D.C. oversaw the project as the team leader, provided strategic guidance on research direction, and supervised the progress of the study. Z.Z. and B.X. provided critical feedback and contributed to the final editing of the manuscript. All authors have read and agreed to the published version of the manuscript.

Funding: This work was supported by the following grants: National Natural Science Foundation of China (Grant No.: 52409120), China Postdoctoral Science Foundation, 75th Batch of General Projects (Grant No.: 2024M752626), Postdoctoral Fellowship Program of CPSF (Grant No.: GZC20232157), and Shaanxi Province Postdoctoral Research Project (Grant No.: 2023BSHEDZZ105).

Data Availability Statement: The data that supports the findings of this study are available from the corresponding author upon reasonable request.

Conflicts of Interest: All authors certify that they have no affiliations with or involvement in any organization or entity with any financial interests or non-financial interests in the subject matter or materials discussed in this manuscript.

References

1. Bao, H.Y.; Long, L.T.; Fu, L.; Wei, J.F. Study on stability of load regulation transition process of hydro turbine governor in power mode. *Trans. Chin. Soc. Agric. Eng.* **2019**, *35*, 50–57, (In Chinese with English abstract). [CrossRef]
2. Binama, M.; Su, W.-T.; Li, X.-B.; Li, F.-C.; Wei, X.-Z.; An, S. Investigation on pump as turbine (PAT) technical aspects for micro hydropower schemes: A state-of-the-art review. *Renew. Sustain. Energy Rev.* **2017**, *79*, 148–179. [CrossRef]
3. Zhang, B.T. Energy Revolution Inevitable, Pumped Storage Thriving. *Hydropower Pumped Storage* **2017**, *3*, 5. Available online: http://old.hydropower.org.cn/showNewsDetail.asp?nsId=22105 (accessed on 13 October 2024). (In Chinese with English abstract).
4. Yang, W.; Norrlund, P.; Chung, C.Y.; Yang, J.D.; Lundin, U. Eigen-analysis of hydraulic-mechanical-electrical coupling mechanism for small signal stability of hydropower plant. *Renew. Energy* **2018**, *115*, 1014–1025. [CrossRef]
5. Khalid, M.; AlMuhaini, M.; Aguilera, R.P.; Savkin, A.V. Method for planning a wind–solar–battery hybrid power plant with optimal generation-demand matching. *IET Renew. Power Gen.* **2018**, *12*, 1800–1806. [CrossRef]
6. Zhang, H.; Hu, P. Study on regulating oscillation in multi hydro turbine generators sharing flow channel structure of tailrace surge tank. *J. Hydraul. Eng.* **2015**, *46*, 10. Available online: http://jhe.ches.org.cn/jhe/ch/reader/view_abstract.aspx?file_no=201502013 (accessed on 1 December 2024). (In Chinese with English abstract).
7. Goyal, R.; Gandhi, B.K. Review of Hydrodynamics Instabilities in Francis Turbine During Off-design and Transient Operations. *Renew. Energy* **2018**, *116*, 697–709. [CrossRef]
8. Li, H.H.; Xu, B.B.; Arzaghi, E.; Abbassi, R.; Chen, D.Y.; Aggidis, G.A.; Zhang, J.J.; Patelli, E. Transient Safety Assessment and Risk Mitigation of a Hydroelectric Generation System. *Energy* **2020**, *196*, 117135. [CrossRef]
9. Liu, Y.N.; Yang, J.D.; Zeng, W. Setting conditions of surge chambers at hydropower station in elastic model based on operation stability. *J. Hydroelectr. Eng.* **2016**, *35*, 6, (In Chinese with English abstract). [CrossRef]
10. Kamal, M.M.; Abbas, A.; Prasad, V.; Kumar, R. A numerical study on the performance characteristics of low head Francis turbine with different turbulence models. *Mater. Today Proc.* **2022**, *49*, 349–353. [CrossRef]
11. El-Sapa, S.; Lotfy, K.; El-Bary, A. Impact of laser short-pulse heating and variable thermal conductivity on photo-elasto-thermodiffusion waves of fractional heat excited semiconductor. *J. Electromagn. Waves Appl.* **2022**, *36*, 2628–2646. [CrossRef]
12. El-Sapa, S.; Lotfy, K.; El-Bary, A.A. Non-local semiconductor photothermal excitation medium in moisture diffusivity subjected to heating by pulsed laser. *Eur. Phys. J. Plus* **2023**, *138*, 368. [CrossRef]
13. El-Sapa, S.; Lotfy, K.; El-Bary, A. Laser short-pulse impact on magneto-photo-thermo-diffusion waves in excited semiconductor medium with fractional heat equation. *Acta Mech.* **2022**, *233*, 3893–3907. [CrossRef]
14. Celebioglu, K.; Altintas, B.; Aradag, S.; Tascioglu, Y. Numerical research of cavitation on Francis turbine runners. *Int. J. Hydrogen Energy* **2017**, *42*, 17771–17781. [CrossRef]
15. Yang, W.; Norrlund, P.; Saarinen, L.; Yang, J.; Guo, W.; Zeng, W. Wear and tear on hydro power turbines—Influence from primary frequency control. *Renew. Energy* **2016**, *87*, 88–95. [CrossRef]
16. Valentín, D.; Presas, A.; Egusquiza, E.; Valero, C.; Egusquiza, M.; Bossio, M. Power Swing Generated in Francis Turbines by Part Load and Overload Instabilities. *Energies* **2017**, *10*, 2124. [CrossRef]
17. Zhang, L.; Wu, Q.Q.; Ma, Z.Y.; Wang, X.N. Transient vibration analysis of unit-plant structure for hydropower station in sudden load increasing process. *Mech. Syst. Signal Process.* **2019**, *120*, 486–504. [CrossRef]
18. Ma, W.C.; Zhao, Z.G.; Yang, J.B.; Lai, X.; Liu, C.P.; Yang, J.D. A transient analysis framework for hydropower generating systems under parameter uncertainty by integrating physics-based and data-driven models. *Energy* **2024**, *297*, 131141. [CrossRef]
19. Zhu, Z.; Tan, X.; Lu, X.; Liu, D.; Li, C. Hopf bifurcation and parameter sensitivity analysis of a doubly-fed variable-speed pumped storage unit. *Energies* **2022**, *15*, 204. [CrossRef]

20. Lei, L.W.; Li, F.; Kheav, K.L.; Jiang, W.; Luo, X.Q.; Patelli, E.; Xu, B.B.; Chen, D.Y. A start-up optimization strategy of a hydroelectric generating system: From a symmetrical structure to asymmetric structure on diversion pipes. *Renew. Energy* **2021**, *180*, 1148–1165. [CrossRef]
21. Li, W.X. Study on the Hydraulic Transition Process Calculation of Three Different Arrangements of Tailwater Pressure Chamber. Master's Thesis, Xi'an University of Technology, Xi'an, China, 2021. (In Chinese with English abstract).
22. Liao, Y.W.; Yang, W.J.; Zhao, Z.G.; Li, X.D.; Ci, X.H.; Bidgoli, M.A.; Yang, J.D. Influence mechanism of backlash nonlinearity on dynamic regulation stability of hydropower units. *Sustain. Energy Technol. Assess.* **2022**, *51*, 101917. [CrossRef]
23. Liu, H.; Fan, B.Y.; Tang, C.; Ge, S.Y.; Wang, Y.; Guo, L. Game theory based alternate optimization between expansion planning of active distribution system and siting and sizing of photovoltaic and energy storage. *Autom. Electr. Power Syst.* **2017**, *41*, 38–45, 116. Available online: https://epjournal.csee.org.cn/dlxtzdh/article/id/564fb3c1-4b09-489c-a3ae-210fd7d344a5 (accessed on 4 December 2024). (In Chinese with English abstract).
24. Singh, M.K.; Naresh, R.; Gupta, D.K. Optimal tuning of temporary droop governor of hydro power plant using genetic algorithm. In Proceedings of the 2013 International Conference on Energy Efficient Technologies for Sustainability, Nagercoil, India, 10–12 April 2013; pp. 1132–1137. [CrossRef]
25. Feng, X.Y. Analysis of the Transient Performance of the Water Turbine Regulation System. Master's Thesis, North China University of Water Resources and Electric Power, Baoding, China, 2016. (In Chinese with English abstract).
26. Riasi, A.; Tazraei, P. Numerical analysis of the hydraulic transient response in the presence of surge tanks and relief valves. *Renew. Energy* **2017**, *107*, 138–146. [CrossRef]
27. Guo, W.C.; Peng, Z.Y. Hydropower system operation stability considering the coupling effect of water potential energy in surge tank and power grid. *Renew. Energy* **2019**, *134*, 846–861. [CrossRef]
28. Hou, J.J.; Li, C.S.; Guo, W.C.; Fu, W.L. Optimal successive start-up strategy of two hydraulic coupling pumped storage units based on multi-objective control. *Int. J. Electr. Power Energy Syst.* **2019**, *111*, 398–410. [CrossRef]

energies

MDPI

Article

Effect of Blade Profile on Flow Characteristics and Efficiency of Cross-Flow Turbines

Ephrem Yohannes Assefa [1,2] and Asfafaw Haileselassie Tesfay [1,2,3,*]

[1] School of Mechanical and Industrial Engineering, Ethiopian Institute of Technology, Mekelle, Mekelle University, Mekelle P.O. Box 231, Ethiopia; ephrem.yohannes@mu.edu.et
[2] Institute of Energy, Mekelle University, Mekelle P.O. Box 231, Ethiopia
[3] Department of Civil Environmental Engineering, Norwegian University of Science and Technology, 7491 Trondheim, Norway
* Correspondence: asfafaw.h.tesfay@ntnu.no (A.H.T.)

Abstract: This study presents a comprehensive numerical investigation into the influence of blade profile geometry on the internal flow dynamics and hydraulic performance of Cross-Flow Turbines (CFTs) under varying runner speeds. Four blade configurations, flat, round, sharp, and aerodynamic, were systematically evaluated using steady-state, two-dimensional Computational Fluid Dynamics (CFD) simulations. The Shear Stress Transport (SST) k–ω turbulence model was employed to resolve the flow separation, recirculation, and turbulence across both energy conversion stages of the turbine. The simulations were performed across runner speeds ranging from 270 to 940 rpm under a constant head of 10 m. The performance metrics, including the torque, hydraulic efficiency, water volume fraction, pressure distribution, and velocity field characteristics, were analyzed in detail. The aerodynamic blade consistently outperformed the other geometries, achieving a peak efficiency of 83.5% at 800 rpm, with improved flow attachment, reduced vortex shedding, and lower exit pressure. Sharp blades also demonstrated competitive efficiency within a narrower optimal speed range. In contrast, the flat and round blades exhibited higher turbulence and recirculation, particularly at off-optimal speeds. The results underscore the pivotal role of blade edge geometry in enhancing energy recovery, suppressing flow instabilities, and optimizing the stage-wise performance in CFTs. These findings offer valuable insights for the design of high-efficiency, site-adapted turbines suitable for micro-hydropower applications.

Keywords: cross-flow turbine; blade profile; hydraulic efficiency; internal flow; CFD and micro-hydropower

1. Introduction

1.1. Background and Motivation

Approximately 700 million people worldwide—primarily in Sub-Saharan Africa and South Asia—remain without access to electricity, according to the International Energy Agency (IEA) [1]. In these regions, the expansion of centralized power grids is often economically and logistically challenging. As a result, decentralized renewable energy systems are being promoted by global frameworks, such as the United Nations Sustainable Development Goal 7 (Affordable and Clean Energy), which emphasizes inclusive and localized energy solutions [2,3].

Micro-hydropower (MHP) has emerged as a particularly attractive option for decentralized electrification due to its low environmental impact, site-specific adaptability, and reliable performance. Among various MHP technologies, Cross-Flow Turbines (CFTs), also known as Michell-Banki turbines, stand out for their ability to operate under low-to-medium heads (2–200 m) [4,5], accommodate flow rates from 0.025 to 13 m^3/s, and tolerate sediment-rich conditions [6]. Their simple mechanical design, ease of local fabrication, and reliable operation with minimal maintenance make them highly suitable for remote and underserved communities [7].

Despite these advantages, the adoption of CFTs has lagged behind conventional turbine technologies, such as Pelton, Francis, and Kaplan, largely due to their comparatively lower hydraulic efficiencies (typically 70–85%, versus 90–95% for conventional designs) and the absence of standardized, performance-optimized design methodologies [8–10]. A key factor influencing the performance of CFTs is the blade geometry, particularly the leading and trailing edge profiles, which play a critical role in the flow interaction, energy transfer, and mechanical durability. The leading edge determines the jet penetration and flow attachment, while the trailing edge governs the flow detachment and wake dynamics [11,12]. Sharp-edged profiles may enhance guidance and reduce separation but are more susceptible to erosion, whereas rounded profiles offer improved structural resilience but may induce turbulence and associated energy losses [13]. Design parameters such as blade curvature, thickness, and inclination angle directly affect both the hydraulic performance and structural integrity [14].

Although considerable research has addressed nozzle configurations, flow regulation strategies, and runner speed optimization, systematic studies on blade profile effects in CFTs remain limited [15–18]. In contrast, conventional turbines have benefited from aerodynamic advancements, including twisted blades in Francis turbines for cavitation control and adjustable blades in Kaplan turbines for load adaptability. Some computational investigations into CFT blade profiles have reported inconsistent performance outcomes under varying flow conditions [19–21]. Moreover, the interaction between blade geometry and the two-phase flow behavior inherent in the open-runner configuration of CFTs remains poorly understood, limiting the generalization of design guidelines.

Existing optimization efforts for CFTs predominantly adopt single-objective approaches focusing solely on the hydraulic efficiency. However, for sustainable deployment in decentralized systems often operating under budgetary, material, and fabrication constraints, a multi-objective design framework is crucial. Advanced blade geometries may yield efficiency gains but frequently require precision manufacturing and specialized materials, thereby increasing costs. Conversely, flat or rounded profiles may be easier to fabricate and structurally more robust, but can underperform in turbulent or unsteady flow conditions. Enhancing the hydraulic efficiency of CFTs through targeted blade design improvements is critical for increasing their cost-effectiveness and competitiveness in decentralized electrification initiatives. Given the operational constraints typical of off-grid and rural applications, such as limited fabrication capabilities and budgetary limitations, optimized blade geometries must provide a balance between performance gains and practical manufacturability.

Furthermore, in the context of decentralized hybrid renewable energy systems, the integration of hydropower with solar photovoltaic (PV) and wind energy technologies is increasingly essential [22]. Such hybrid configurations improve overall system performance, reduce environmental impact, enhance cost efficiency, and strengthen long-term resilience. Therefore, the development of efficient and adaptable CFTs directly supports the broader goals of sustainability, reliability, and universal energy access in emerging and underserved regions.

Therefore, an effective design must balance hydraulic performance, structural durability, and manufacturability.

Computational Fluid Dynamics (CFD) provides a high-resolution, non-invasive method for analyzing internal flow phenomena, including jet impingement dynamics, velocity distributions, pressure gradients, and turbulence development, across the two energy conversion stages. Among the turbulence models, the Shear Stress Transport (SST) k–ω model is widely favored for its robust near-wall treatment and capacity to predict flow separation. Nonetheless, its limitations in modeling free shear layers warrant consideration of more advanced models such as Scale-Adaptive Simulation (SAS) and Detached Eddy Simulation (DES), which provide improved turbulence resolution at higher computational costs [23,24]. For validation, experimental techniques such as Particle Image Velocimetry (PIV) and Laser Doppler Anemometry (LDA) offer precise, high-resolution flow field measurements and are essential for validating numerical predictions [25].

This study aims to systematically investigate the influence of blade profile geometry on the internal flow characteristics and hydraulic performance of Cross-Flow Turbines using CFD-based numerical simulations. Four geometries, flat, round, sharp, and aerodynamic, were evaluated under a constant head of 10 m and varying runner speeds (270 to 940 rpm). The analysis focuses on the velocity field evolution, pressure distribution, and turbulence intensity during both energy conversion stages. Key performance metrics include stage-wise efficiency and energy losses attributed to jet impingement, leakage, and air-water entrainment. The findings are expected to guide the development of more efficient, durable, and cost-effective CFT designs for decentralized hydropower applications.

1.2. Principles of Flow Characteristics and Operation

The CFT facilitates energy conversion through a two-stage process that capitalizes on both the reaction and impulse mechanisms (see Figure 1). In the inward flow first stage (reaction-dominated), water enters the turbine through a nozzle where its pressure energy is converted into kinetic energy. The high-velocity jet impinges tangentially on the runner blades, initiating rotation. As the water flows radially inward, guided along the curved blades, it experiences a pressure drop that induces a reaction force, contributing significantly to the torque generation. As the water flows through the central air-filled region of the runner—maintained at atmospheric pressure—the flow undergoes a directional shift. In the outward flow second stage (impulse-dominated), the remaining kinetic energy of the fluid is extracted as the water is redirected outward across the blade surface.

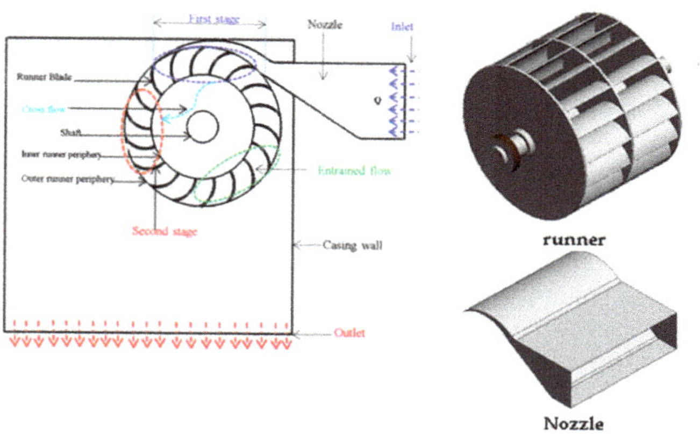

Figure 1. Flow path characteristics and components of the CFT.

This secondary interaction enables additional energy recovery through the impulse action, while the open-center configuration ensures negligible backpressure and supports efficient pressure recovery. A central element governing this two-stage energy extraction is the blade geometry, which directly affects the flow guidance, pressure distribution, and momentum transfer. The blade curvature, along with the inlet (β_1) and outlet (β_2) angles, defines the velocity triangles that govern the tangential component of the fluid velocity responsible for torque production. The optimal blade design minimizes flow separation, reduces turbulence intensity, and ensures high hydraulic efficiency throughout the passage. The runner configuration consists of radially arranged blades enclosed between two circular end discs, forming a cylindrical rotor that accommodates both energy conversion stages under varying hydraulic loads. High-fidelity CFD analysis, supplemented by empirical design principles, was employed to characterize the internal flow behavior and optimize the geometric parameters for improved performance under a specified head (H) and flow rate (Q). This dual approach ensures that the design refinements align with both theoretical expectations and practical operational constraints.

1.3. Turbine Configuration and Design Approach

1.3.1. Nozzle Design

The CFT receives water from the penstock, where it is accelerated through a converging nozzle to form a high-velocity jet that impinges on the runner blades. The exit velocity of this jet, which is critical for performance modeling, is estimated using Bernoulli's principle accounting for head losses:

$$V_1 = C_n\sqrt{2gH} \tag{1}$$

where C_n is the nozzle loss coefficient (0.95), g is the gravitational acceleration, and H is the net head.

The nozzle is angularly positioned to span with a nozzle entry angle of 90° (ranging between 40° and 130°), ensuring effective flow admission to the first runner stage (Figure 2a).

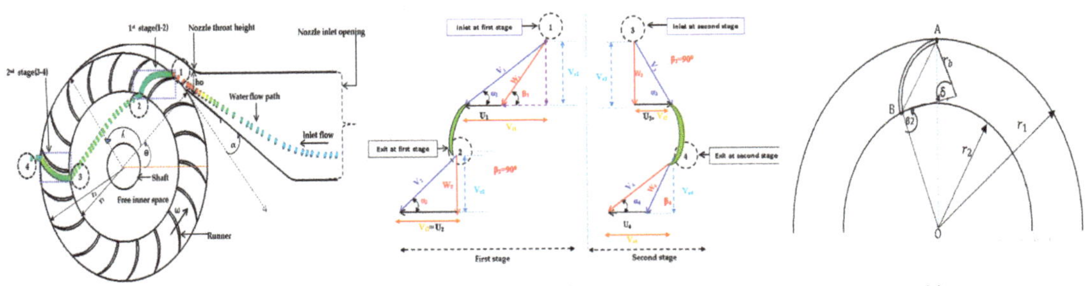

(a) (b) (c)

Figure 2. (**a**) Flow path and design workflow; (**b**) Velocity triangles and (**c**) Blade parameters.

To quantify the inlet conditions accurately using the numerical method, the area-weighted velocity magnitudes and directions were extracted from a cross-sectional plane intersecting the nozzle arc. As illustrated in Figure 2b, the tangential component of the velocity is then expressed as:

$$V_t = V\cos\alpha_1 \tag{2}$$

where $\alpha_1 = 16°$ is the jet incidence angle optimized for effective entry alignment.

The relative velocity angle at the blade entry (β_1) is then computed as follows:

$$\tan(\beta_1) = 2\tan(\alpha_1) \tag{3}$$

The fluid advances through the first-stage blades, and assuming ideal flow conditions, the shock losses at the entry of the second stage are considered insignificant. A 90° (β_2) inter-blade angle is incorporated between the stages to ensure smooth and efficient flow redirection.

1.3.2. Operating Parameter

The blade peripheral velocity, relative to the jet speed, defines the velocity ratio:

$$V_r = \frac{U_1}{V_1} = \frac{1}{2}\cos \propto_1 \tag{4}$$

An optimal velocity ratio of approximately 0.48 was selected to maximize the energy transfer to the runner. Using this, the runner diameter is calculated from:

$$D_1 = \frac{42.3\cos(\propto_1)\sqrt{H_1}}{N} \tag{5}$$

1.3.3. Blade Geometry and Runner Design

The blade design parameters, including the number, spacing, and curvature, are derived empirically and geometrically. The jet thickness is expressed as: $S_1 = kD_1$, where k (empirical coefficient) is often K = 0.085 for practical cases.

Blade spacing (t_1), number of blades (Z_b), and runner width (B_w) are calculated respectively by:

$$t_1 = \frac{S_1}{\sin \beta_1} \tag{6}$$

$$Z_b = \frac{\pi D_1}{t_1} \tag{7}$$

$$B_w = \frac{Q}{S_1 * V_1} \tag{8}$$

In CFT, the aspect ratio (a_r), the ratio of the runner diameter (D) to blade width (b), often termed the runner length, exhibits a fundamental geometric and functional interdependence. This relationship directly affects the turbine's ability to accommodate flow, maintain structural stability, and achieve optimal hydraulic performance, making it a critical consideration in turbine design. The Aspect ratio (a_r) must be between 0.8 and 1.2 for optimal flow and structural integrity [26].

The diameter ratio, $D_r = 0.66$ was selected to optimize the flow recovery in the second stage. The blade curvature radius (r_b) and central angle (δ) are then derived on the basis of the runner geometry and velocity triangle (see Figure 2c) as follows:

$$r_b = \frac{D_1\left(1 - D_r^2\right)}{4\cos \beta_1} \tag{9}$$

$$\tan\left(\frac{\delta}{2}\right) = \frac{\cos \beta_1}{\sin \beta_1 + D_r} \tag{10}$$

1.3.4. Performance Evaluation

Hydraulic efficiency is defined by the ratio of the mechanical power to the available water power:

$$\eta_h = \frac{\sum T * \omega}{\gamma * H * Q} \tag{11}$$

where T is the torque and ω is angular velocity. The torque is evaluated via numerical post-processing in ANSYS CFX® 2023 as:

$$T = \left[\int r * (\bar{\bar{\tau}}.n))ds)\right].a \tag{12}$$

where, ds is the position vector, $\overline{\overline{\tau}}$ is the stress tensor, n is the unit normal to the surface, and a is aligned with the rotational axis. Circular sampling planes were positioned along the runner perimeters to extract the velocity and pressure profiles at both stages, facilitating a comprehensive assessment of the flow behavior and efficiency trends.

2. Methodology

2.1. Blade Profile and Turbine Specifications

In this study, four blade profiles such as flat, round, sharp edged, and aerodynamic, were selected based on their manufacturability and anticipated hydraulic performance (see Figure 3). Each blade design was geometrically modeled with uniform thickness, except for the sharp-edged profile, which featured a 1 mm leading edge, and the aerodynamic profile, which adopted an asymmetric airfoil-inspired geometry to improve flow adherence and minimize separation.

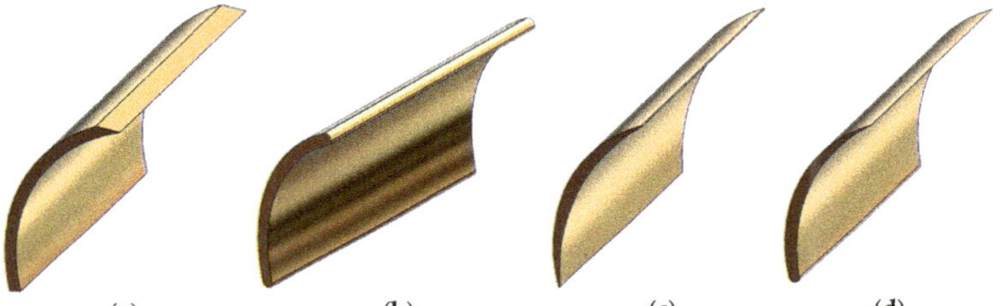

Figure 3. Blade profiles considered: (**a**) Flat, (**b**) Round, (**c**) Sharp edged, and (**d**) Aerodynamic.

The turbine geometry was designed to reflect a typical micro-hydropower CFT configuration. Table 1 summarizes the key design parameters, which remained constant across all simulations to isolate the influence of the blade profile variations.

Table 1. Geometrical and operating specifications of the CFT used in the simulation.

Design Parameter	Symbol	Value	Unit
Effective pressure head	H	10	m
Flow rate	Q	55	l/s
Flow attack angle	α_1	16	deg
Blade entry angle (1st stage)	β_1	30	deg
Blade exit angle (1st stage)	β_2	90	deg
Optimal runner speed	N	670	rpm
Outer runner diameter	D_1	200	mm
Inner runner diameter	D_2	130	mm
Diameter ratio	D_r	0.66	-
Jet thickness	S_1	17	mm
Blade spacing	t_1	34	mm
Number of blades	Z_b	18	-
Blade thickness (except sharp)	t	3	mm
Nozzle entry arch angle	λ	90	deg
Blade curvature radius	r_b	33	mm
Central angle	δ	73.5	deg
Blade width	B_w	225	mm

2.2. Computational Setup

The computational domain was modeled using the ANSYS Design Modeler and included three main components: the nozzle, the runner with blades, and the casing (see Figure 4). To accurately capture the interaction between the stationary and rotating regions, the Multiple Reference Frame (MRF) approach was employed. The runner and blades were assigned to the rotating zone, while the nozzle and casing were treated as stationary.

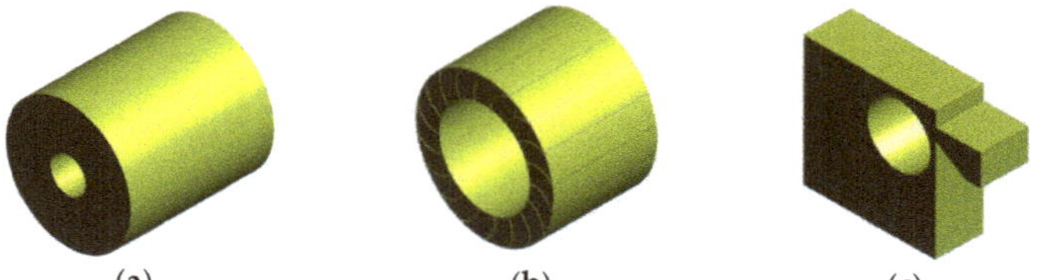

(a) (b) (c)

Figure 4. Computational domain components: (**a**) Free inner zone, (**b**) Runner, and (**c**) Nozzle and casing.

To model the interaction between the stationary and rotating zones, a frozen rotor interface was implemented. This approach retains the relative orientation between frames, facilitating a quasi-steady-state simulation of the flow while avoiding the high computational demand associated with fully transient simulations. Although three-dimensional modeling offers enhanced capability in resolving secondary flows and complex turbulence structures, prior studies have shown that 3D simulations often report slightly lower efficiency predictions by approximately 3% to 4%, due to additional secondary flow losses [27,28]. Given the comparative nature of this study and its focus on isolating the effects of blade geometry, a two-dimensional steady-state simulation provides an optimal balance between computational efficiency and accuracy.

2.3. Mesh Sensitivity and Grid Convergence Study

To ensure the reliability and numerical accuracy of the CFD simulations, a systematic grid sensitivity analysis was conducted, followed by a quantitative uncertainty evaluation using the Grid Convergence Index (GCI) method. The computational domain—including runner, nozzle, and casing—was discretized with a structured tetrahedral mesh (See Figure 5). Critical regions such as blade surfaces and boundary layers were refined using a local element size of 1 mm, incorporating 20 inflation layers with a growth rate of 1.2, resulting in wall-resolved y^+ values below 5, thus meeting the requirements of near-wall turbulence modeling.

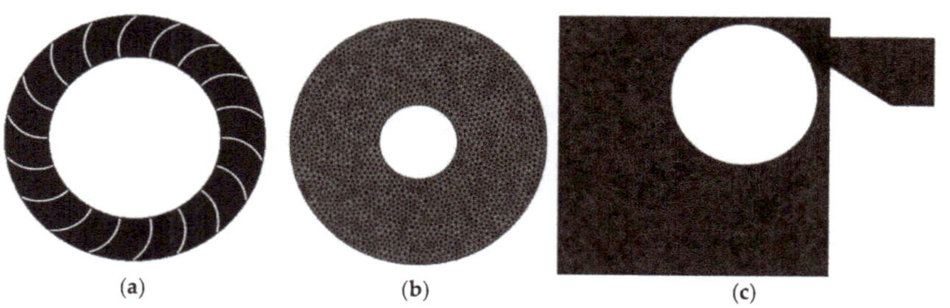

(a) (b) (c)

Figure 5. Discretization of the computational domain: (**a**) Runner domain, (**b**) Free inner surface, and (**c**) Casing and nozzle regions.

An initial mesh independence test was conducted by varying the characteristic element size from 10 mm to 1 mm. Solution stability, particularly in the predicted torque and mass flow rate, was achieved below the 1 mm threshold, which informed subsequent grid refinement levels (see Figure 6).

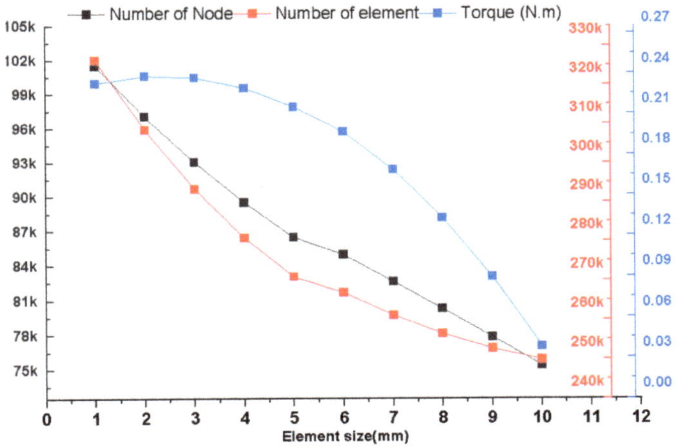

Figure 6. Grid sensitivity analysis based on the torque convergence.

Three progressively refined grids were generated with characteristic element sizes of 1.0 mm (coarse), 0.75 mm (medium), and 0.5 mm (fine), as shown in Table 2. Torque (τ) was selected as the primary quantity of interest. The refinement ratios between the mesh levels were calculated as $r_{21} = \frac{h_2}{h_1}$ and $r_{32} = \frac{h_3}{h_2}$, Where h_1,h_2, and h_3 are the characteristic grid spacings of the fine, medium, and coarse meshes, respectively.

Table 2. Torque results across the grid levels.

Grid Level	Number of Cells (N)	Torque (Nm)	Grid Spacing (h)
Coarse	321,500	0.243	0.00176
Medium	620,000	0.2435	0.00127
Fine	915,600	0.2438	0.00105

The observed order of convergence (q) was determined via Richardson extrapolation [29,30], and the theoretical torque was estimated using standard formulations (Equation (14)).

$$q_{n+1} = \frac{\ln\left[\left\{\frac{\tau_3-\tau_2}{\tau_2-1}\left(r_{12}{}^{q_n}-1\right)+r_{12}{}^{q_n}\right\}\right]}{\ln\left(\frac{r_{12}}{r_{23}}\right)} \tag{13}$$

$$\tau_{exact} = \tau_1 - \left(\frac{\tau_2-\tau_1}{r_{12}{}^{q_{n+1}}-1}\right) \tag{14}$$

The GCI was computed for each mesh transition using a refinement factor and safety factor of 1.25 as follows.

$$GCI_{12} = F_s\left[\frac{1}{\tau_1} * \frac{\tau_2-\tau_1}{r_{12}{}^{qn}-1}\right] * 100\% \tag{15}$$

$$GCI_{32} = F_s\left[\frac{1}{\tau_2} * \frac{\tau_2-\tau_3}{r_{23}{}^{qn}-1}\right] * 100\% \tag{16}$$

The relative errors between the grid levels were calculated as

$$\varepsilon_{21} = \left|\frac{\tau_1-\tau_2}{\tau_1}\right| \tag{17}$$

$$\varepsilon_{32} = \left| \frac{\tau_2 - \tau_3}{\tau_2} \right| \tag{18}$$

where

- ε_{ij} is the relative error between the mesh level i (coarse) and j (finer)
- τ_j torque on the coarse grid
- τ_i torque on the finer grid

The results (see Table 3) show that relative errors and GCI values remained within acceptable engineering limits (GCI < 2%), with the observed order of accuracy q = 0.33.

Table 3. Grid convergence and numerical uncertainty metrics.

Parameters	Symbol	Value	Unit
Fine-to-medium refinement ratio	r_{21}	1.39	-
Medium-to-coarse refinement ratio	r_{32}	1.22	-
Observed order of accuracy	q_{n+1}	0.33	-
Richardson's extrapolated torque	τ_{exact}	0.25	N.m
Relative Error (fine-to-medium)	(ε_{21})	0.0013	-
Relative Error (medium-to-coarse)	(ε_{32})	0.0014	-
Grid Convergence Index (fine-to-medium)	GCI_21	1.38	(%)
Grid Convergence Index (medium-to-coarse)	GCI_32	1.56	(%)

Consequently, the 1-mm mesh was selected for all subsequent simulations, balancing computational efficiency and prediction accuracy. This mesh resolution adequately captured critical flow features such as separation, reattachment, and wake development in the turbine domain.

2.4. Boundary Conditions and Turbulence Modeling

The computational domain was initialized with appropriate boundary conditions to replicate the realistic operating conditions of the CFT. A constant pressure inlet boundary condition was applied to simulate a net hydraulic head of 10 m, corresponding to a water volume fraction (α-$_w$) of 1 and an air fraction (α-$_a$) of 0. At the outlet, an atmospheric pressure condition was imposed with and, allowing free discharge of the flow. No-slip, adiabatic wall conditions were imposed on all wall boundaries, including the casing, runner, and blades, to accurately model the viscous effects and thermal insulation. The interface between the stationary and rotating regions was handled using the frozen rotor approach within the Multiple Reference Frame (MRF) framework. This method enables the simulation of the steady-state rotor–stator interaction with reduced computational expense compared to the transient rotor–stator models. Turbulence effects were modeled using the Shear Stress Transport (SST) model, which is known for its robustness in handling complex internal flows. The SST model combines the near-wall accuracy of the formulation with the free-stream stability of the model through a blending function. This hybrid approach is well-suited for resolving the boundary layer separation, recirculation zones, and shear-dominated regions prevalent in turbine applications.

2.5. Numerical Solution and Governing Equations

Steady-state simulations were carried out using a homogeneous, two-phase, free-surface model in ANSYS CFX, treating water and air as interpenetrating continua with shared velocity, pressure, and turbulence fields. This approach simplifies the interface treatment while preserving the essential dynamics of the free-surface flow relevant to CFT operations. The governing equations are based on the Reynolds-Averaged Navier–Stokes (RANS) formulation, modified to include rotational effects in the rotating domain via a rotating reference frame [31,32]. These equations consist of:

- Continuity equation (mass conservation):

$$\frac{\partial(\alpha_P\rho_P)}{\partial x} + \nabla.(\alpha_P\rho_P) = S_P \tag{19}$$

- Momentum conservation (in rotating frame)

$$\begin{aligned}\frac{\partial\left(\alpha_P\rho_P\vec{V}_r\right)}{\partial t} + \nabla.\left(\alpha_P\rho_P\vec{V}_r*\vec{V}_r\right) - \nabla.\left(\mu_{eff}\nabla\vec{V}_r\right) + \alpha_P\rho_P(2\vec{\omega}\times\vec{v}_r + \vec{\omega}\times\vec{\omega}\times\vec{r} \\ + \vec{\alpha}\times\vec{r} + \vec{a}) = \nabla\dot{P} + \nabla.\left(\mu_{eff}\nabla\vec{V}_r\right)^T + S_{MP}\end{aligned} \tag{20}$$

where

$$\dot{P} = P + \frac{2}{3}\rho_P k + \frac{2}{3}\mu_{eff}\nabla.\vec{V} \tag{21}$$

The variable α_P represents the phase volume fraction, while α_P denotes the phase density, and S_P indicates the phase mass flow rate. $\vec{V}_r$ is the relative flow velocity, $\vec{\omega}$ is the angular velocity, and $2\vec{\omega}\times\vec{v}_r$ r represents the Coriolis acceleration. The term $\vec{\omega}\times\vec{\omega}\times\vec{r}$ accounts for the centripetal acceleration, and, $\vec{\alpha}\times\vec{r}$ reflects the acceleration due to irregular variations in the rotational speed, alongside $\vec{a}$ linear changes in the relative velocity. In all cases, the subscript "p" is used to refer to the specific properties of each phase.

- Effective viscosity using the SST model

$$\mu_{eff} = \mu + \mu_t \tag{22}$$

where μ (kg/m·s) is the dynamic viscosity, and μ_t (kg/m·s) represents the turbulent viscosity. In the k-ε model, the turbulent viscosity is calculated using Equation (23), which depends on the turbulent kinetic energy (k) and the dissipation rate (ε):

$$\mu_t = \frac{C_\mu \rho k^2}{\varepsilon} \tag{23}$$

where $C\mu$ (-) is a constant, ρ (kg/m³) is the fluid density, k (J/kg) is the turbulent kinetic energy, ε (m²/s³) is the turbulent dissipation rate, and ω (s^{-1}) is the mean turbulent frequency. In contrast, the k-ω model computes the turbulent viscosity using the turbulent kinetic energy and the turbulent frequency (ω) as shown in Equation (24).

$$\mu_t = \rho\frac{k}{\omega} \tag{24}$$

- Homogeneous two-phase mixture properties

The mixture density and viscosity were calculated based on the volume fractions of each phase as: [11,32].

$$\rho = \sum_P \alpha_P\rho_P \tag{25}$$

$$\mu = \sum_{p} \alpha_p \mu_p \quad (26)$$

where α (-) is the volume fraction, p represents the phase, and ρ (kg/m³) and μ (kg/m·s) are the density and viscosity, respectively, of the average mixture used in the modified RANS equations. The air density ρa (kg/m³) is assumed to be a function of the pressure P (Pa) according to the state equation:

2.6. Simulation Scenarios and Performance Evaluation

To evaluate the influence of blade geometry on turbine performance, four blade profiles, such as flat, round, sharp, and aerodynamic, were systematically investigated under varying runner speeds of 270, 540, 670, 800, and 940 rpm.

These speeds represent the operational ranges commonly encountered in MHP applications. The primary performance indicators considered were torque, hydraulic efficiency, and internal flow behavior. For each scenario, the simulations assessed the velocity ratio and its relationship with both global (overall) and stage-wise hydraulic performance. A detailed grid sensitivity analysis was performed for all configurations, and the Grid Convergence Index (GCI) was computed to ensure the numerical accuracy and reliability of the results. Post-processing focused on analyzing the velocity and pressure fields to capture internal flow phenomena, including flow separation, recirculation zones, and vortex structures. Special attention was directed toward the interaction between the incoming jet and blade surfaces, particularly the leading and trailing edge impingement zones, as these regions are critical in determining the energy transfer efficiency across both turbine stages.

3. Results and Discussion

This section presents a comprehensive analysis of the velocity and pressure distributions across four blade profiles—flat, round, sharp, and aerodynamic —under varying runner speeds (270–940 rpm), highlighting their influence on torque generation, flow dynamics, and energy conversion efficiency.

3.1. Velocity Field and Flow Component Analysis

The velocity field within the runner was examined to evaluate the impact of the blade profile and runner speed on the internal flow behavior and turbine performance. Figure 7 compares the relative velocity fields across the flat, round, sharp, and aerodynamic blades at various runner speeds.

At 270 rpm (Figure 8), the flat and round blades exhibited high relative velocities and prominent flow misalignment, leading to turbulence and poor flow attachment near the blade leading edges. In contrast, the sharp and aerodynamic blades demonstrated improved flow guidance and reduced separation, with the aerodynamic profile showing the most coherent velocity patterns.

At 670 rpm (Figure 9), a more favorable alignment between the blade motion and the incoming flow was observed. The aerodynamic blade profile facilitates smooth velocity transitions with minimal flow disturbance, contributing to the peak hydraulic efficiency. Sharp blades also maintain effective flow control, whereas flat and round blades continue to suffer from localized recirculation.

At 940 rpm (Figure 10), the relative velocities declined across all profiles due to the reduced velocity differentials. However, the aerodynamic blade preserves effective flow control and minimizes losses. Sharp blades show moderate performance, while flat and round blades display intensified turbulence and vortex shedding near the trailing edge.

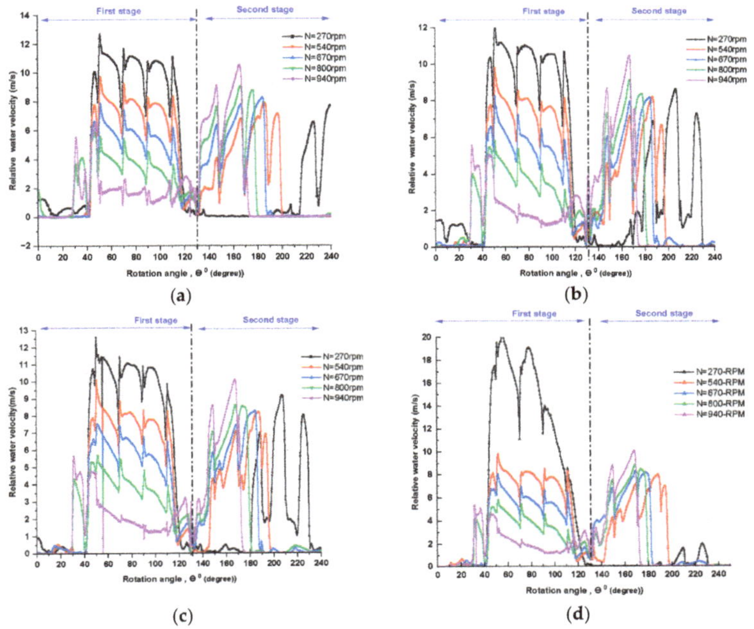

Figure 7. Effect of runner speed and blade profile on the relative velocity at the outer runner periphery: (**a**) Flat, (**b**) Round, (**c**) Sharp, and (**d**) Aerodynamic.

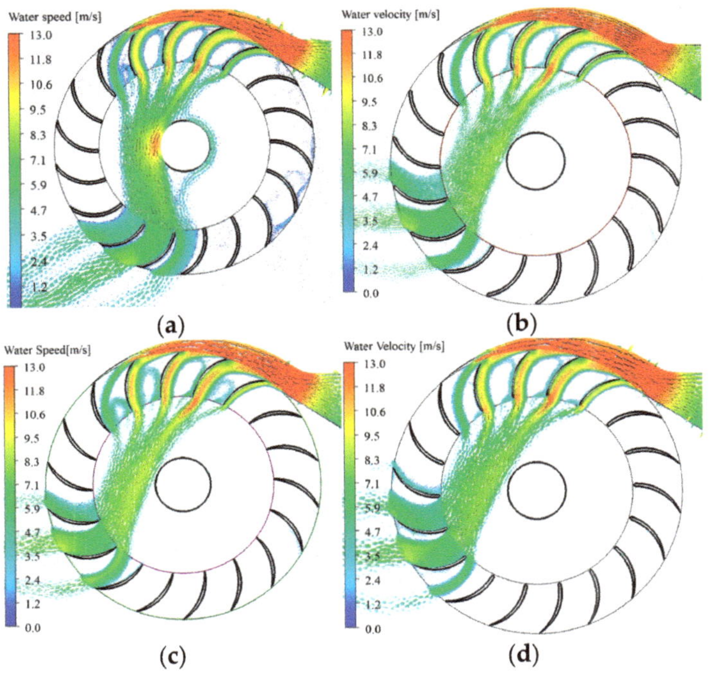

Figure 8. Water velocity distribution for different blade profiles at 270 rpm: (**a**) Flat, (**b**) Round, (**c**) Sharp, and (**d**) Aerodynamic.

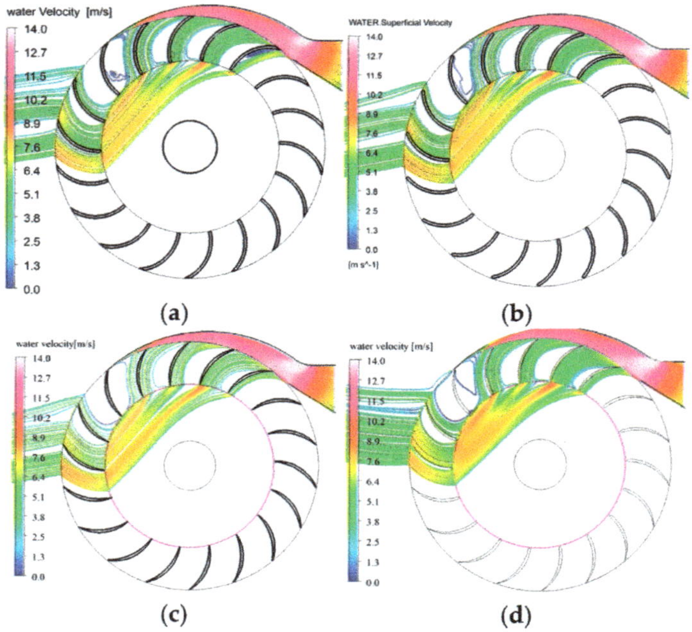

Figure 9. Water velocity distribution profiles at 670 rpm: (**a**) Flat, (**b**) Round, (**c**) Sharp, and (**d**) Aerodynamic.

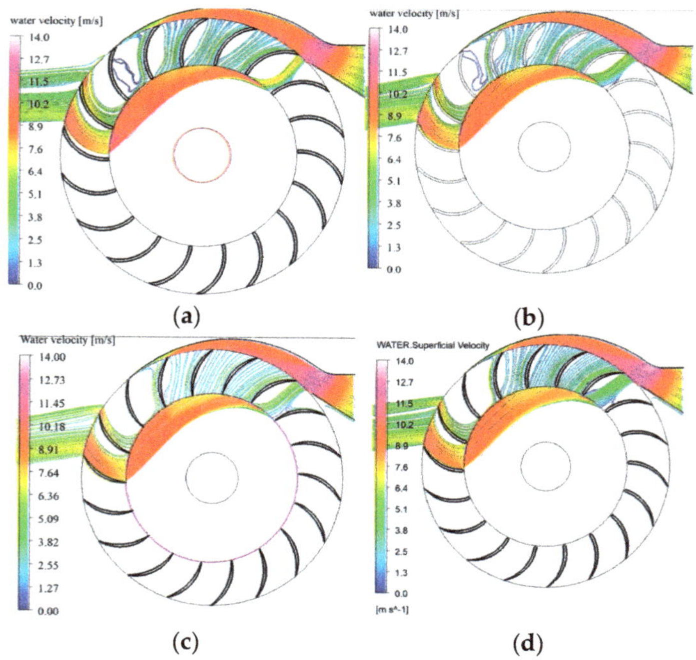

Figure 10. Water velocity distribution profiles at 940 rpm: (**a**) Flat, (**b**) Round, (**c**) Sharp, and (**d**) Aerodynamic.

3.1.1. Tangential Velocity Component

Figure 11 illustrates the distribution of the tangential velocity at the runner periphery. At low speeds (270 rpm), a significant portion of the tangential momentum exits the runner unutilized, especially for flat and round blades. As the speed increases to 670 rpm, the aerodynamic and sharp blades demonstrate improved deceleration of the tangential velocity, enhancing the torque generation. At speeds beyond 800 rpm, the reduction in the tangential velocity differential limits further energy extraction. Nonetheless, the aerodynamic blades maintained better momentum transfer and reduced losses compared to the other profiles.

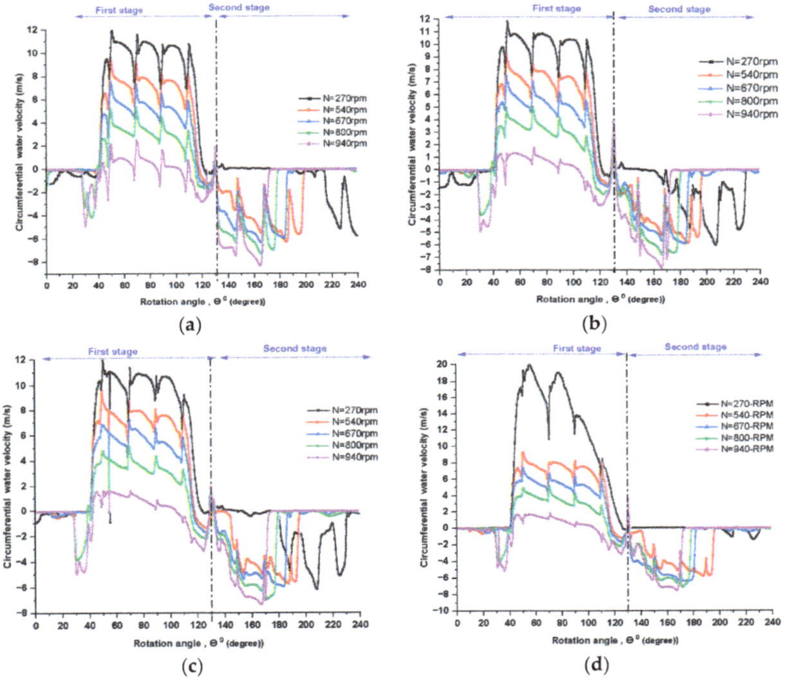

Figure 11. Effect of runner speed and blade profile on the tangential velocity component at the outer runner periphery: (**a**) Flat, (**b**) Round, (**c**) Sharp, and (**d**) Aerodynamic.

3.1.2. Radial Velocity Component

The radial velocity distributions at 800 rpm are shown in Figure 12. The flat and sharp blades exhibit flow irregularities, including backflow and radial dispersion. The aerodynamic and round blades maintain a smoother radial progression and reduced separation. The aerodynamic blade, in particular, sustains a more uniform radial flow conducive to stable energy transfer. In general, Aerodynamic and sharp blades consistently promote favorable flow structures across runner speeds. Peak performance is observed at 670–800 rpm, where both tangential and radial velocity components are optimally aligned with the blade motion. At lower and higher speeds, increased turbulence and misalignment reduce the performance, particularly in flat and round profiles. Among the tested geometries, the aerodynamic blade provided the most stable flow, lowest turbulence, and highest efficiency across the operational range.

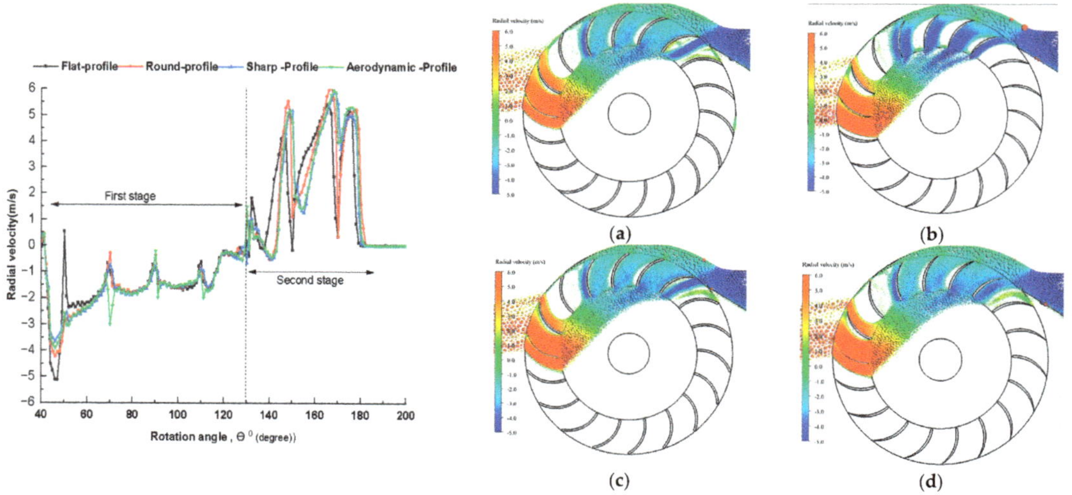

Figure 12. Effect of blade profile on radial flow distribution at 800 rpm: (**a**) Flat, (**b**) Round, (**c**) Sharp, and (**d**) Aerodynamic.

3.2. Pressure Distribution Analysis

The pressure distribution within the runner domain provides critical insight into the energy conversion mechanisms across the blade profiles and operating speeds. Figure 13 presents the static pressure contours for different blade geometries and runner speeds. As expected, the pressure gradients intensified with increasing runner speed, with the most significant static pressure drop occurring in the first stage, confirming its partially reactive behavior. In contrast, the second stage—vented to atmospheric conditions—exhibits the characteristics of an impulse stage.

At lower speeds (270–540 rpm), inefficient blade–fluid interaction leads to non-uniform pressure fields and stagnation zones, particularly with flat and round blades. These profiles suffer from premature pressure losses due to poor flow redirection and localized recirculation (Figure 14). Aerodynamic blades demonstrate a more favorable pressure recovery, with smoother gradients across the runner.

At the optimal runner speed of 670 rpm, the aerodynamic profile exhibits the steepest and most uniform pressure drop (Figure 15), indicating efficient energy transfer and minimal flow detachment. Sharp blades also maintain a relatively smooth pressure profile, whereas flat and round blades show residual pockets of high pressure due to flow misalignment.

At higher speeds (800–940 rpm), increased blade–flow velocity differentials induce localized pressure disturbances, especially for flat and round blades. The aerodynamic and sharp blades mitigate these instabilities by maintaining smooth deceleration and pressure recovery. Figure 16 details the angular variation in the static and total pressures along the runner periphery at 800 rpm. The aerodynamic blade sustains a gradual pressure decline through the first stage and prevents abrupt pressure recovery in the second, thereby reducing the cavitation risks.

Overall, the pressure distribution trends confirm that the blade geometry significantly influences the internal flow stability and energy extraction. Aerodynamic blades consistently exhibit superior pressure gradients and smoother transitions, especially at optimal runner speeds, reinforcing their suitability for high-efficiency CFT.

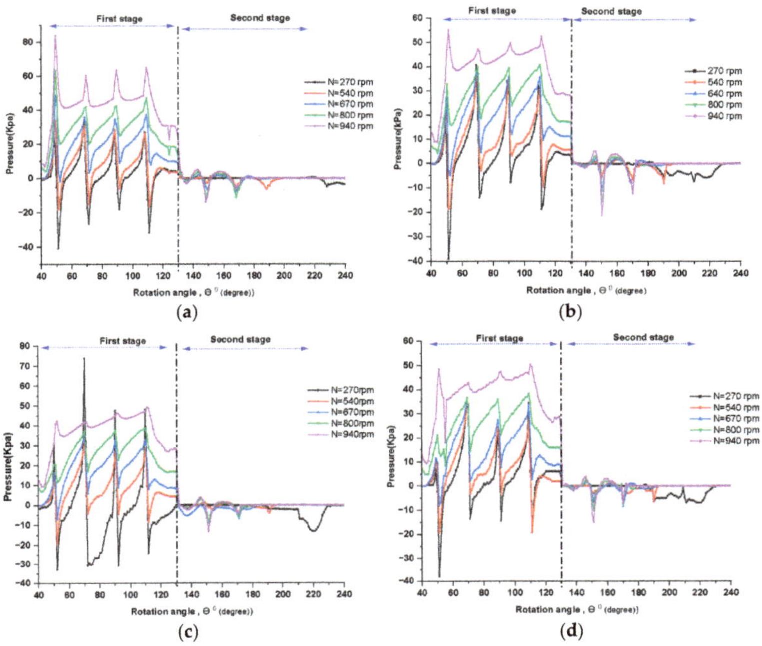

Figure 13. Static Pressure distribution profile at different runner speeds: (**a**) Flat, (**b**) Round, (**c**) Sharp, and (**d**) Aerodynamic.

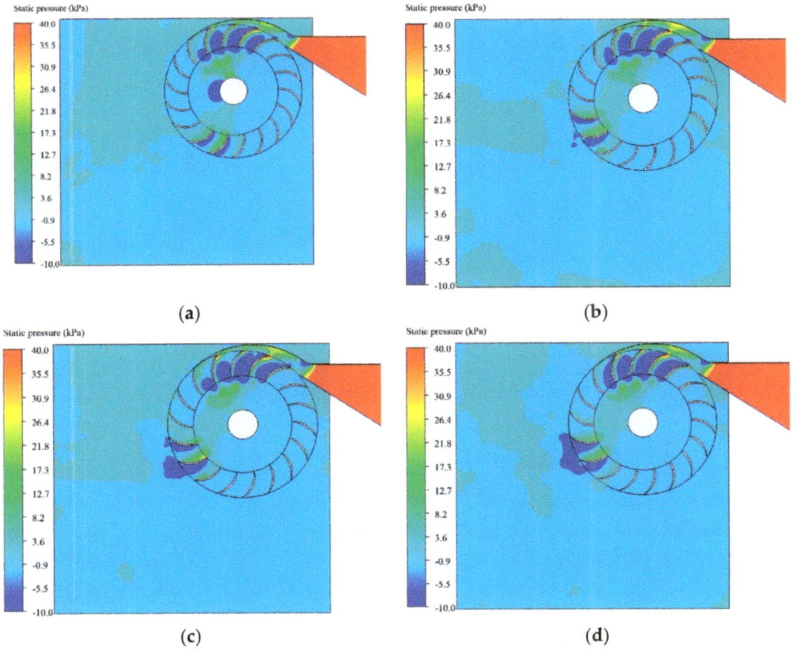

Figure 14. Static pressure distribution profile at 540 rpm: (**a**) Flat, (**b**) Round, (**c**) Sharp, and (**d**) Aerodynamic.

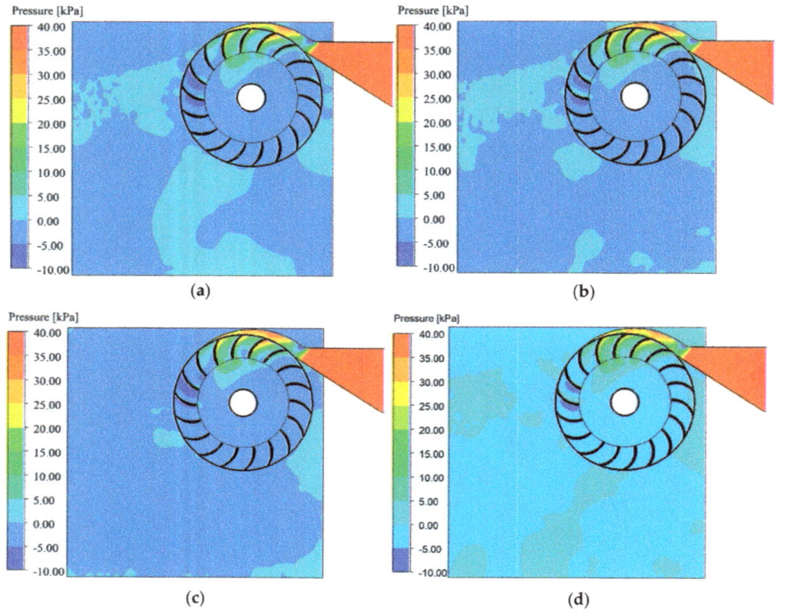

Figure 15. Static pressure distribution at 670 rpm for the four blade profiles: (**a**) flat, (**b**) round, (**c**) Sharp, and (**d**) Aerodynamic.

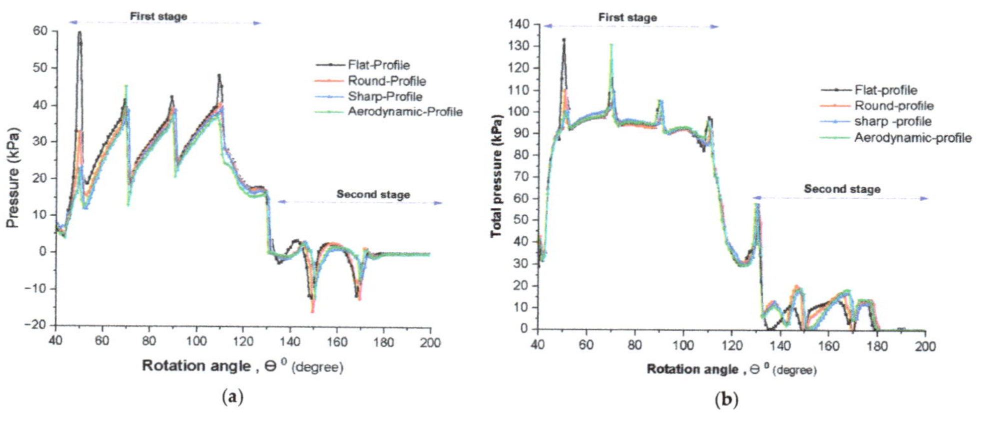

Figure 16. Pressure distribution profile at 800 rpm for the four blade profiles: (**a**) static pressure and (**b**) Total pressure.

3.3. Water Volume Fraction Distribution

The water volume fraction (WVF), representing the proportion of the liquid phase within the computational domain, serves as a critical metric for evaluating the flow stability, phase continuity, and potential cavitation within a CFT. A near-unity WVF indicates minimal entrainment of vapor or air and confirms consistent energy transmission across the rotor stages. Figure 17 depicts the water fraction profile at the optimal runner speed. The aerodynamic blade demonstrates the most favorable performance, sustaining high liquid phase continuity across both turbine stages. In contrast, the flat and round blades showed localized voids near the trailing edge and hub regions, indicative of poor flow guidance and incipient cavitation.

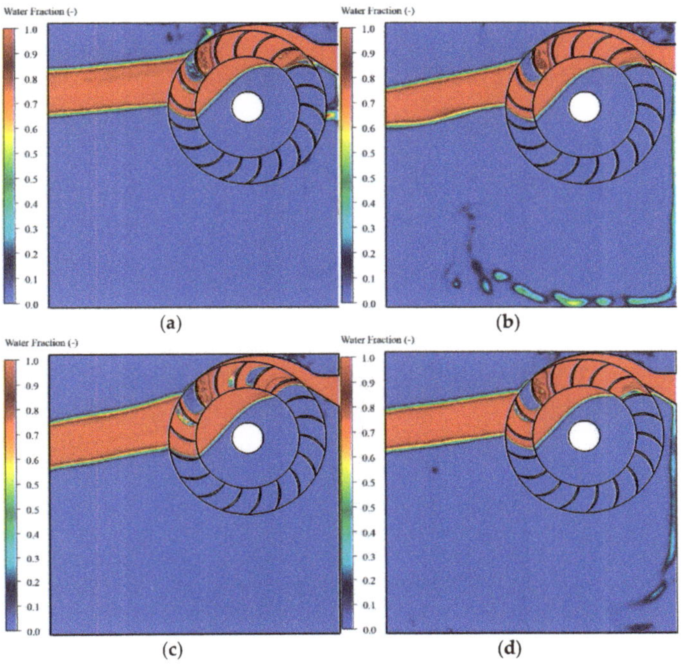

Figure 17. Effect of WVF distribution profile at 800 rpm for the four blade profiles: (**a**) Flat, (**b**) Round, (**c**) Sharp, and (**d**) Aerodynamic.

At low runner speeds (e.g., 270 rpm), reduced blade rotation relative to the flow velocity leads to significant flow misalignment, vortex shedding, and recirculation. These effects result in sporadic vapor pockets near structural boundaries—particularly in the flat and round blade configurations—due to their suboptimal hydrodynamic profiles. As the runner speed increases to intermediate levels (540–670 rpm), the blade motion becomes better synchronized with the incoming flow, thereby reducing void formation and promoting full-phase continuity.

At the optimal speed of 670 rpm, the WVF nears unity across the entire runner for the sharp and aerodynamic blade profiles, reflecting improved flow attachment, minimal detachment zones, and reduced turbulence. These profiles facilitate smoother transitions and preserve water integrity throughout the rotor, especially in the second stage, where pressure recovery is critical. However, at higher runner speeds (800–940 rpm), increased blade-induced turbulence introduces secondary flow instabilities, particularly in the flat and sharp profiles. Despite this, the aerodynamic blades continue to exhibit the most stable WVF distribution, sustaining near-complete water occupancy throughout the domain. Figure 18 further illustrates the localized WVF variations at 800 rpm, emphasizing the enhanced flow continuity achieved by the aerodynamic profile.

3.4. Influence of Runner Speed and Blade Geometry on the Internal Flow Behavior

A precise understanding of the internal flow behavior in CFTs is vital for optimizing blade performance under various operating conditions. Critical flow phenomena such as separation, recirculation, vortex shedding, and pressure recovery directly impact turbine efficiency and are highly dependent on both runner speed and blade geometry.

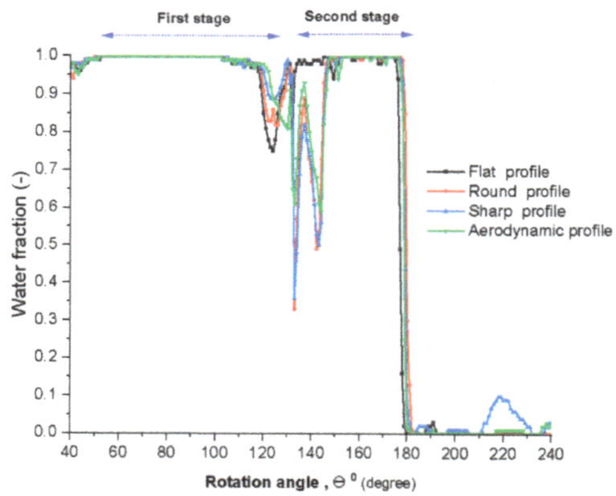

Figure 18. Localized water volume fraction distribution at 800 rpm for different blade profiles.

3.4.1. Flat Blade Profile

Flat blades, characterized by their abrupt leading and trailing edges, introduce significant hydrodynamic disturbances. At low runner speeds (270 rpm), the inflowing jet encounters sudden deflection surfaces, triggering early flow separation and large recirculation zones downstream (see Figure 19-a). With increasing speeds (540–670 rpm), unsteady vortex shedding at the trailing edge intensifies turbulence, reducing momentum transfer.

Figure 19. Streamline evolution and recirculation patterns for the flat blades at 270, 670, 800 and 940 rpm.

With increasing speeds (540–670 rpm), unsteady vortex shedding at the trailing edge intensifies turbulence, reducing momentum transfer. At high speeds (800–940 rpm), adverse pressure gradients and amplified centrifugal forces further destabilize the flow, leading to considerable efficiency loss

3.4.2. Round Blade Profile

Round blades, with their smooth curvature, enhance flow continuity by minimizing abrupt transitions. At low speeds (270–540 rpm), the gradual turning of the flow reduces stagnation and detachment compared to flat blades (see Figure 20).

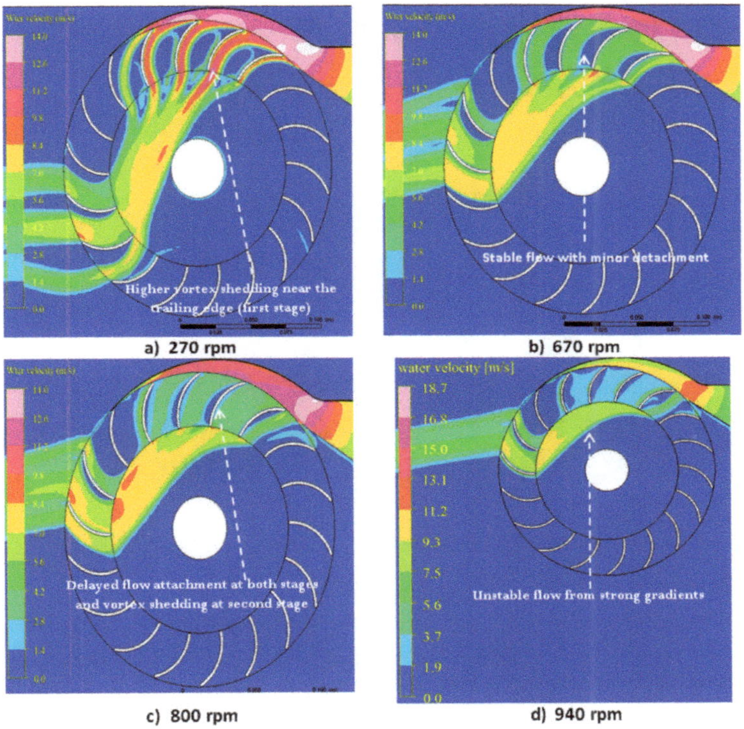

Figure 20. Streamline evolution for the round blades at 270, 670, and 940 rpm.

While moderate recirculation zones persist near the trailing edge, turbulence remains subdued. At higher speeds (800–940 rpm), coherent vertical structures reappear due to steeper velocity gradients, though they are notably weaker than those seen with flat profiles.

3.4.3. Sharp-Edged Blade Profile

Sharp-edged blades are tailored for high flow deflection and efficient energy transfer within a narrow speed window. At intermediate runner speeds (540–670 rpm), they demonstrate effective jet redirection with limited separation (see Figure 21). However, at lower speeds (270 rpm), insufficient flow momentum causes premature separation and strong vortex formation near the blade root. At higher speeds (800–940 rpm), intense pressure gradients at the trailing edge promote vigorous vortex shedding, resulting in substantial energy dissipation.

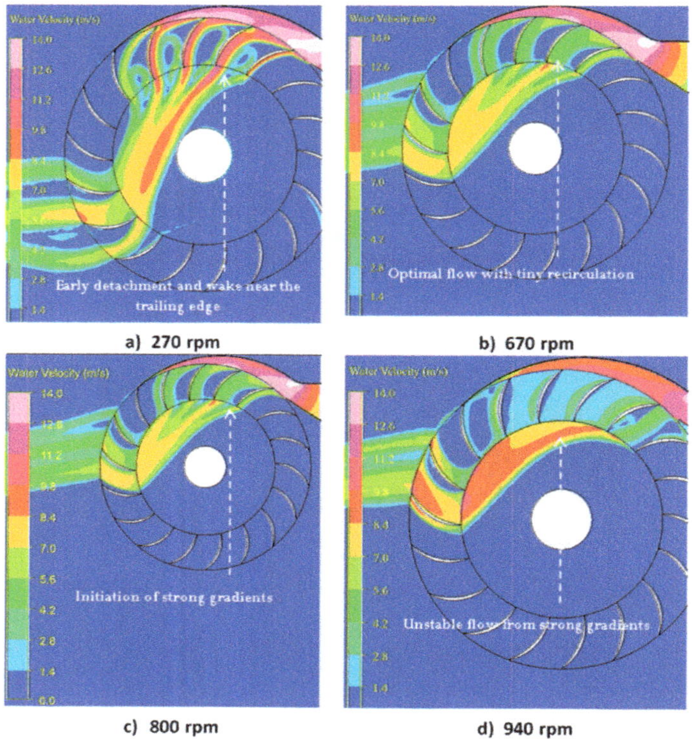

Figure 21. Streamline evolution for the sharp-edged blades at 270, 670, and 940 rpm.

3.4.4. Aerodynamic Blade Profile

The aerodynamic blade, modeled on a NACA airfoil, exhibits superior flow behavior across all tested runner speeds (Figure 22). At 270 rpm, the smoothly contoured profile maintains attached flow with negligible wake formation. Between 540 and 800 rpm, efficient pressure recovery and momentum transfer are sustained with delayed separation and low turbulence levels. Even at 940 rpm, the flow remains stable and coherent, underscoring the profile's robustness across a wide operational range.

A comparative overview of the internal flow behavior and corresponding hydraulic performance for each blade type across different runner speeds is presented in Table 4.

Table 4. Summarizes the internal flow behavior across the runner speed range for each blade geometry.

Blade Profile	Low Runner Speed (270–540 rpm)	Medium Runner Speed (540–640 rpm)	Higher Runner Speed (800–940 rpm)	Hydraulic Performance
Flat	High separation, strong recirculation	Vortex shedding, unstable wake	Severe instability, low efficiency	Poor
Round	Improved flow, minor voids	Stable flow with minor detachment	Coherent vortex structure	Moderate
sharp	Early detachment, vortex near the root	Optimal flow redirection	Unstable from strong gradients	Speed dependent
Aerodynamic	Attached flow, minimal wake	Stable, efficient pressure recovery	Delayed separation, reduced shedding	High efficiency across most speed ranges

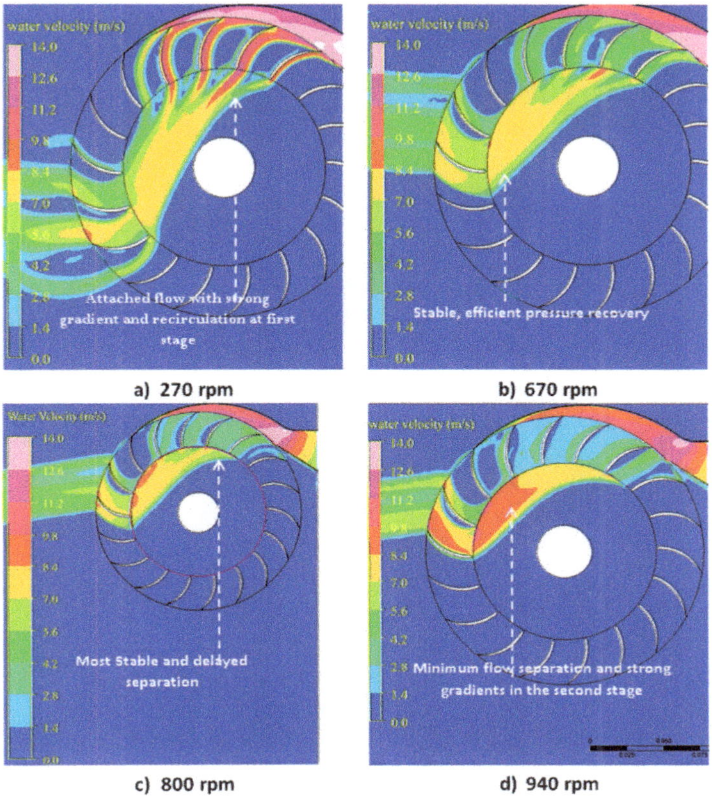

Figure 22. Streamline evolution for the aerodynamic blades at 270, 670, and 940 rpm.

The table highlights that flat and sharp-edged profiles exhibit substantial flow separation and turbulence, particularly outside their optimal speed ranges. In contrast, round and aerodynamic profiles maintain more stable internal flows, resulting in better efficiency and broader operational flexibility.

3.5. Comparative Hydraulic Performance Study

The hydraulic efficiency of the CFT was evaluated across four distinct blade profiles—flat, round, sharp, and aerodynamic—under varying runner speeds corresponding to different velocity ratios. The results reveal a clear dependence of the turbine performance on both the blade geometry and the operational velocity ratio. Among the profiles tested, the aerodynamic and sharp blade geometries demonstrated superior hydraulic performance, particularly at higher velocity ratios. The aerodynamic blade profile achieved the highest hydraulic efficiency of 83.04% at a velocity ratio of 0.84 (runner speed of 800 rpm), closely followed by the sharp blade profile, which reached an efficiency of 83.01% at a velocity ratio of 1.14 (see Figure 23 – c&d). This indicates that these profiles are more effective in directing and maintaining the kinetic energy of the water jet through the runner passage, resulting in reduced energy losses due to flow separation and turbulence.

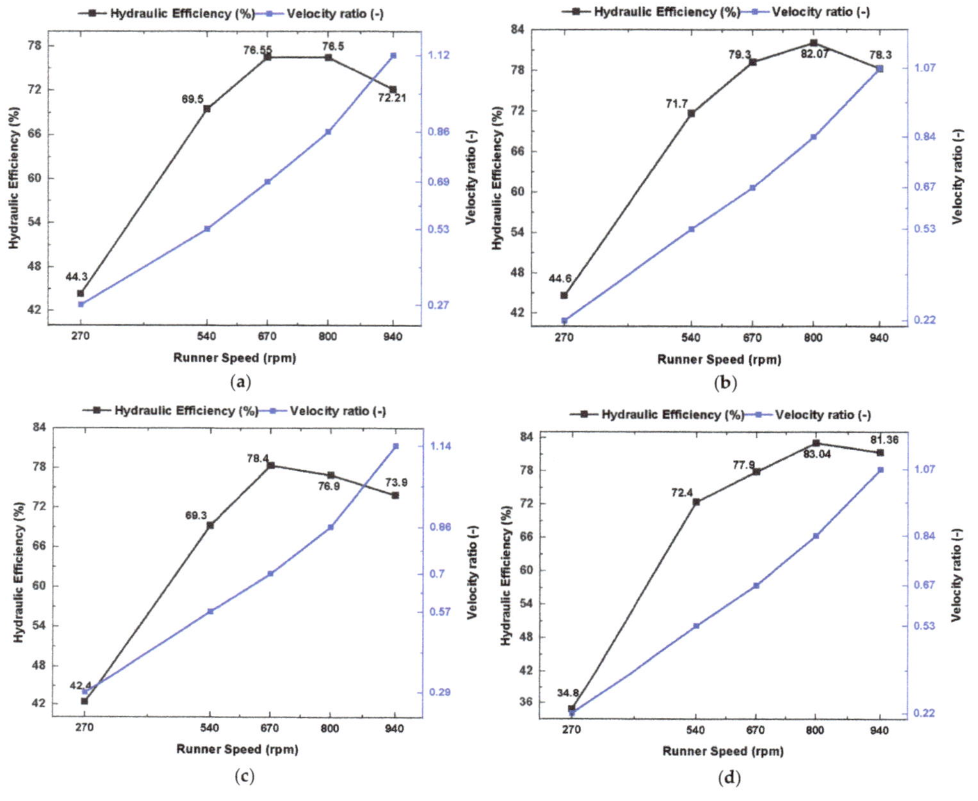

Figure 23. Hydraulic efficiency trends across varying runner speeds and velocity ratio (**a**) Flat, (**b**) Round, (**c**) Sharp, and (**d**) Aerodynamic.

The sharp blade profile consistently outperformed the flat and round profiles across the examined range, achieving over 70% efficiency beyond a velocity ratio of 0.57. This trend suggests that the sharp edge facilitates smoother water entry and reduced flow blockage, which contributes to better energy conversion. Similarly, the aerodynamic blade showed a marked improvement in efficiency with increasing runner speed, beginning from a relatively low efficiency of 34.8% at 270 rpm (velocity ratio 0.22) and progressively reaching 81.36% at 940 rpm (velocity ratio 1.07). This profile appears particularly advantageous in high-speed operation, where streamlined geometry minimizes the drag forces and enhances the flow alignment with the blade curvature. Conversely, the flat and round blade profiles exhibited moderate performance. The flat blade attained a peak efficiency of 76.55% at a velocity ratio of 0.69, while the round blade reached a slightly higher peak of 79.15% at a velocity ratio of 0.69 (See Figure 24). However, both profiles showed a decline in efficiency beyond this point, indicating performance limitations due to flow detachment and recirculation zones at higher operating speeds.

In summary, the results emphasize the critical influence of blade geometry on the hydraulic performance of the CFT. While traditional flat and round profiles provide satisfactory performance within a narrow operational window, sharp and aerodynamic blade designs offer superior and more consistent efficiency across a broader range of velocity ratios. These findings support the adoption of optimized blade profiles—particularly sharp and aerodynamic for enhanced energy capture in MHP applications.

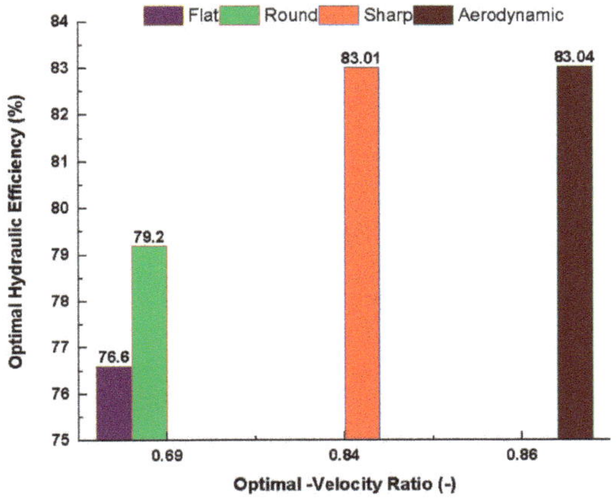

Figure 24. Optimal velocity ratio for each blade profile.

3.6. Torque and Power Output

The torque generation behavior of the CFT was systematically assessed for four blade profiles—flat, round, sharp, and aerodynamic—across a range of runner speeds and corresponding velocity ratios. The analysis reveals important insights into how blade geometry influences the turbine's mechanical response under different operational conditions. For all blade profiles, the measured torque exhibited a declining trend with increasing runner speed (and velocity ratio). This inverse relationship reflects the fundamental principle of turbine operation: as the rotational speed of the runner increases, the angular acceleration rises, while the torque tends to decrease due to the diminishing resistance from the water jet and the reduced effective momentum transfer at higher angular velocities.

At lower speeds (270 rpm), all blade types generated relatively high torque, ranging between 0.34 and 0.36 N·m. The flat blade produced the highest initial torque of 0.36 N·m, closely followed by the aerodynamic and sharp blades at 0.35 N·m (See Figure 25–(a)). This indicates that at lower velocity ratios (0.22–0.29), the kinetic energy of the jet is effectively utilized by most profiles to produce torque, owing to the longer residence time of water within the runner and favorable jet-blade interaction.

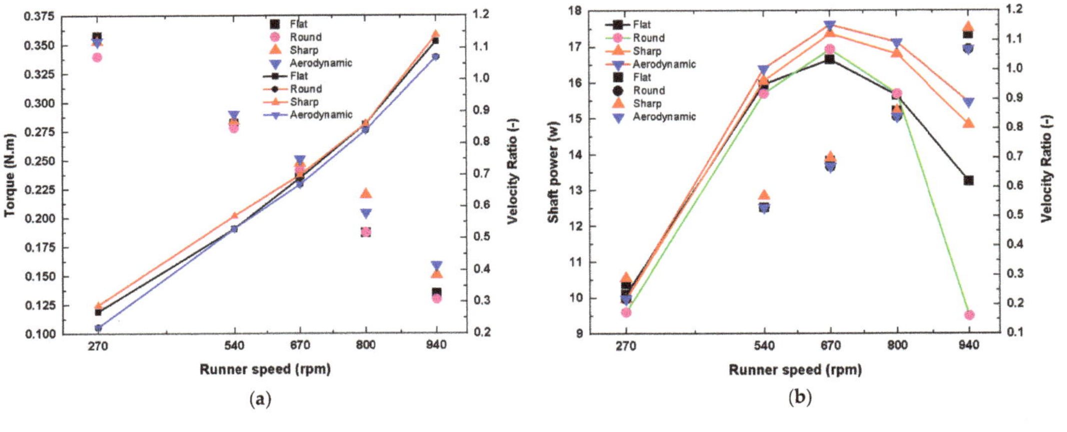

Figure 25. Effect of runner speed across each blade profile. (**a**) Torque generation, (**b**) shaft power.

However, as the runner speed increased, the aerodynamic and sharp blade profiles demonstrated slightly better retention of torque relative to their flat and round counterparts. For instance, at 800 rpm (velocity ratio ≈ 0.84–0.86), the aerodynamic and sharp blades produced torque values of 0.20 N·m and 0.22 N·m, respectively, compared to 0.19 N·m for both the flat and round blades (See Figure 25-(b)). Notably, at the highest speed tested (940 rpm), the sharp and aerodynamic profiles continued to sustain relatively higher torque values (0.15 and 0.16 N·m, respectively), suggesting their enhanced capability to maintain effective jet momentum capture even at elevated angular velocities.

The torque trends further highlight the aerodynamic blade's superior ability to maintain torque generation, particularly in the mid-range velocity ratios (0.53–0.67), where it consistently outperformed others with a peak of 0.29 N·m at a velocity ratio of 0.53. This is likely attributed to its streamlined geometry, which reduces drag and improves flow attachment, thereby facilitating more efficient energy transfer to the rotational motion. In contrast, the flat and round blade profiles displayed the steepest decline in torque, dropping from 0.36 N·m and 0.34 N·m at 270 rpm to just 0.13 N·m at 940 rpm, indicating that their relatively blunt geometry may not sustain optimal jet interaction and energy transfer at higher operational speeds. In summary, while all blade profiles showed similar torque behavior at low speeds, the sharp and aerodynamic blades provided a more favorable torque response across a wider range of velocity ratios. These findings underscore the advantage of the refined blade geometry in sustaining the torque generation efficiency in CFT, particularly under high-speed conditions relevant to practical MHP applications.

3.7. *Stage-Wise Efficiency Analysis*

The CFT operates with a unique two-stage energy conversion mechanism: the first stage, where the jet initially impinges on the runner blades, and the second stage, where the residual kinetic energy is extracted as the flow re-engages the runner on the opposite side. The effectiveness of each stage is strongly governed by the runner speed, blade geometry, and the internal flow dynamics between stages.

3.7.1. First Stage: Primary Energy Extraction Zone

The first stage is the principal energy conversion zone, where the high-velocity jet directly transfers its momentum to the blades, generating torque and accounting for approximately 60–70% of the total hydraulic efficiency. The magnitude of the pressure and velocity gradients is highest in this phase, especially near the blade leading edge. At intermediate runner speeds (540–670 rpm), the round and aerodynamic blades exhibit optimal flow alignment, with reduced vortex formation and efficient pressure recovery. Notably, at 670 rpm, these profiles achieved minimal flow separation, promoting smooth transitions and enhancing energy extraction. In contrast, at low speeds (270 rpm), inadequate blade-jet synchronization causes flow misalignment, recirculation, and premature detachment, especially in flat and sharp-edged profiles. At high runner speeds (800–940 rpm), centrifugal effects intensify, often leading to premature water ejection and disrupted flow paths, thereby diminishing the contribution from the first stage, particularly in non-streamlined blade configurations.

3.7.2. Second Stage: Secondary Energy Recovery Zone

The second stage used the residual kinetic energy from the first-stage outflow. Its effectiveness is highly dependent on the quality of the flow handed over from Stage I. Turbulence, misalignment, or large-scale vortices from the primary stage significantly reduce the recovery potential. The aerodynamic and round blades demonstrate superior second-stage performance, especially at 540–670 rpm, characterized by stable radial velocity fields and high water volume fractions. The flat blades, however, consistently underperform due to persistent separation and poorly structured wakes. At 670 rpm, the second-stage energy contribution peaks at approximately 30–35% for the round and aerodynamic profiles, while the flat blades fall short, exhibiting high turbulence intensity and pressure losses. The findings emphasize the necessity of maintaining coherent flow transitions between stages and achieving high overall efficiency.

3.8. Influence of Blade Geometry on Stage-Wise Efficiency

The blade geometry plays a pivotal role in defining the stage-wise efficiency through its control over the flow deflection, pressure recovery, and turbulence mitigation. As shown in Figure 26, the aerodynamic blades provide the most balanced and robust performance across both stages, especially at higher runner speeds.

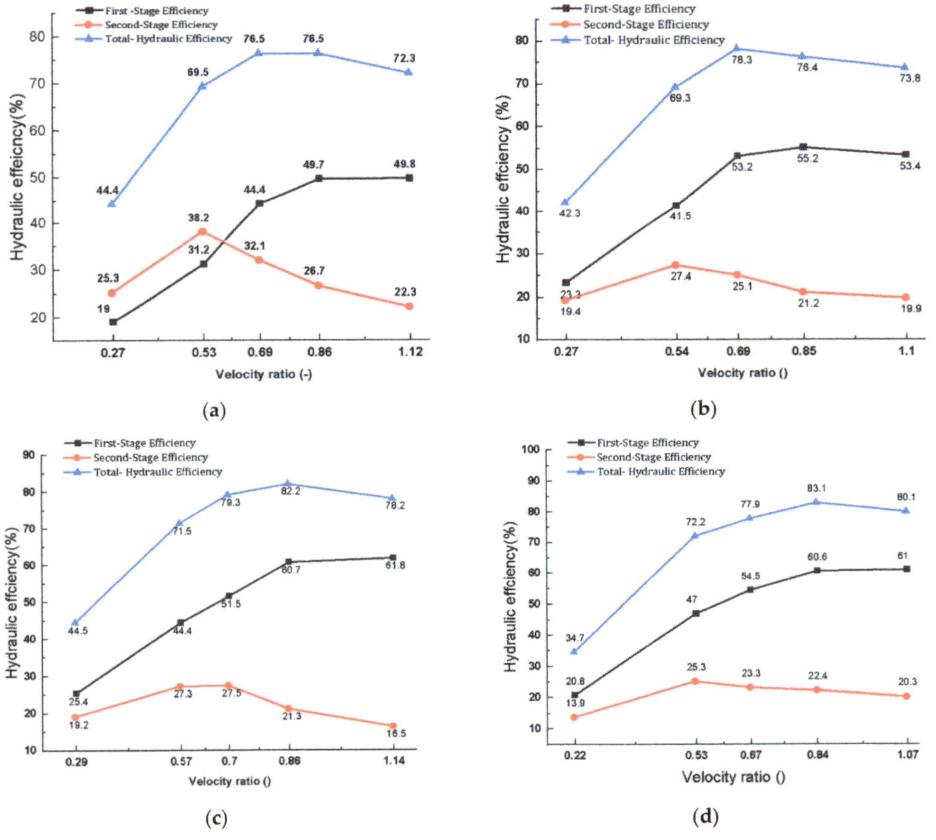

Figure 26. Stage-wise efficiency distribution for (**a**) Flat, (**b**) Round, (**c**) Sharp, and (**d**) Aerodynamic blade profiles at different runner speeds.

- Flat blades show a higher second-stage contribution at lower speeds (e.g., 38% at 540 rpm), but their performance degrades as the speed increases. At 940 rpm, the first stage dominates with 50%, yet the overall efficiency remains modest due to the severe turbulence.
- The round blades deliver peak first-stage efficiency (~53%) at 670 rpm, while the second-stage efficiency peaks (~27%) at 540 rpm, reflecting relatively balanced yet moderate performance.

- ☞ Sharp blades reach maximum first-stage efficiency (62%) at 940 rpm and highest second-stage contribution (~27.5%) at 670 rpm, but exhibit sensitivity to off-optimal conditions.
- ☞ Aerodynamic blades consistently yield the best results, achieving up to 83.1% total efficiency at 940 rpm, with over 70% of the energy extracted during the first stage. At 540 rpm, these blades attain an overall efficiency of 72.2%, with a well-balanced stage-wise contribution.

These results confirm that while flat and sharp-edged blades suffer from inefficient energy distribution and turbulence, round and aerodynamic profiles ensure smoother flow behavior, especially under mid-to-high runner speeds. For optimized CFT performance, the blade geometries must facilitate pressure recovery and minimize flow disruption in both energy conversion stages.

3.9. Exit Pressure Trends and Blade Geometry Influence

Efficient energy extraction in CFTs is closely tied to exit pressure behavior, which serves as an indicator of how effectively the hydraulic energy is converted into mechanical work. Ideally, exit pressures should remain near atmospheric levels sufficiently low to confirm efficient energy recovery, but not so low as to induce cavitation risks. Conversely, elevated exit pressures often signify incomplete kinetic energy extraction and turbulent flow detachment at the runner exit. Among the blade geometries tested, the aerodynamic blades consistently produced the lowest exit pressures, coupled with high residual velocities and minimal turbulence. These flow characteristics favor smoother interactions with downstream components such as draft tubes or tailrace channels, thereby enhancing the overall system efficiency.

Conversely, the flat blades exhibited elevated exit pressures, often exceeding 14 kPa at high speeds, reflecting their limited ability to manage flow detachment and suppress vortex shedding at the exit region.

- ☞ At 270 rpm, the aerodynamic blades achieved the lowest exit pressure (6.62 kPa) with moderate efficiency (44.6%), while the sharp blades, despite a higher exit pressure (9.98 kPa), delivered the highest efficiency in this speed regime (Figure 27). The round blades exhibited a moderate pressure level (7.95 kPa) but yielded the lowest efficiency (42.4%).
- ☞ At 540 rpm, all blade types experienced a drop in the exit pressure. The aerodynamic blades reached peak efficiency (72%), demonstrating superior jet deflection and energy recovery potential. Round and sharp blades followed closely, while flat blades continued to underperform.
- ☞ At 670 rpm, the sharp blades achieved maximum efficiency (79%), albeit with higher exit pressures, suggesting a trade-off between pressure buildup and momentum transfer. Aerodynamic blades maintained high efficiency while also minimizing pressure spikes, indicating robust flow control and favorable blade-jet interaction.
- ☞ At higher runner speeds (800–940 rpm), exit pressures significantly increased across all profiles, with flat blades peaking at ~15 kPa, corresponding to the lowest efficiency (76.5%) in this regime. The aerodynamic blades retained their dominance, maintaining exit pressures below 10 kPa and achieving efficiencies exceeding 83%, attributed to their streamlined profile and reduced wake turbulence.

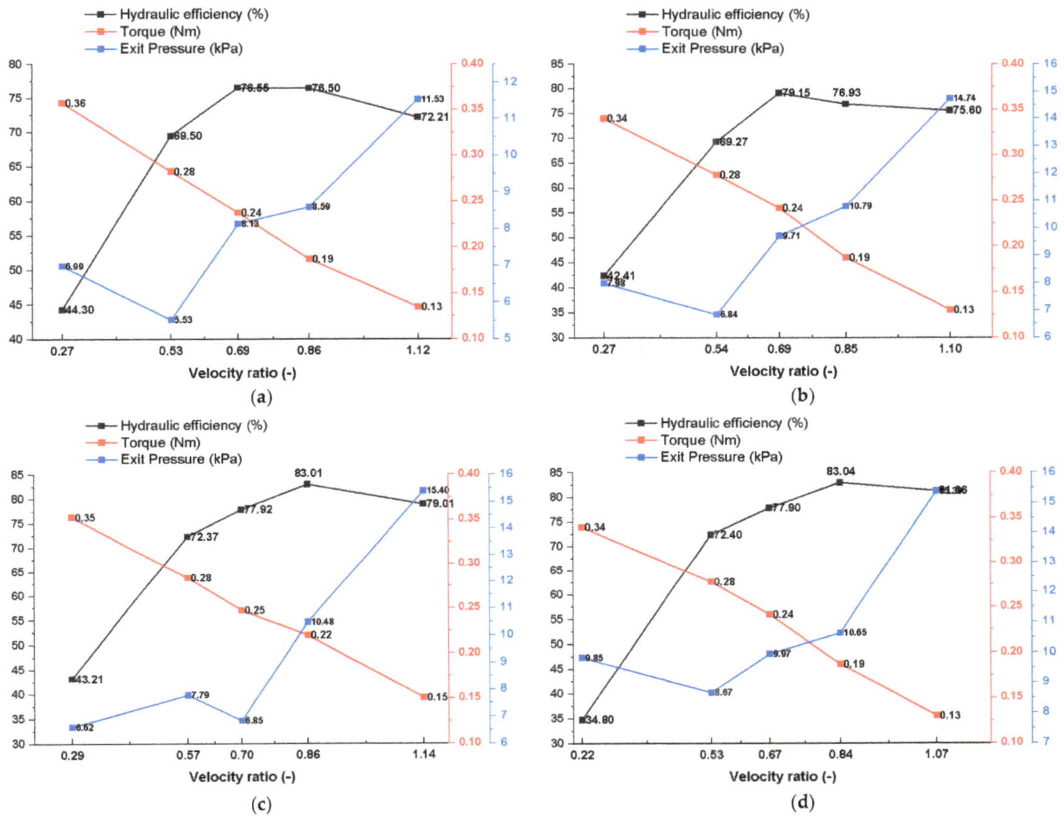

Figure 27. Comparative influence of blade profile on hydraulic efficiency, torque output, and exit pressure distribution: (**a**) Flat, (**b**) Round, (**c**) Sharp, (**d**) Aerodynamic.

- The flat blades exhibited consistently high exit pressures and low performance, highlighting ineffective flow redirection.
- The round blades maintained moderate pressure and efficiency but lacked peak performance.
- Sharp blades delivered high efficiency at select speeds, although they were susceptible to pressure spikes at higher rms.
- Aerodynamic blades outperformed across all conditions, showing low exit pressures and consistent energy recovery, establishing them as the most suitable profile for performance-critical MHP applications.

In summary, the blade geometry governs the exit pressure behavior and post-runner flow quality, which directly affect the stage efficiency and downstream hydraulic interactions. The aerodynamic profile demonstrated superior performance by maintaining favorable pressure conditions, ensuring high energy conversion and operational reliability.

3.10. Comparison with Prior Studies

To contextualize the findings, the performance of the tested blade profiles was compared with findings reported in previous numerical investigations by Naseem [21], Asif [33], and others(see Figure 28). These studies primarily focused on lower runner speed regimes (100–386 rpm), limiting their applicability to high-speed MHP operations. Naseem [21] analyzed the impact of various leading-edge blade profiles across runner speeds ranging from 100 to 240 rpm in 20 rpm increments. The round blade profile achieved the highest efficiency (68%) at 180 rpm, whereas the flat blade consistently underperformed, recording a minimum efficiency of 59%. The aerodynamic and sharp blades in Naseem's work demonstrated near-optimal performance around 160 rpm but showed limited gains with increased speed. Similarly, Asif reported a peak efficiency of 59% for the round blade at 386 rpm. Although both studies revealed consistent performance trends across blade types, the overall efficiency levels remained significantly lower than those obtained in this study. In a related investigation, a modified NACA 6512 airfoil with adjusted curvature and chord length was tested against a conventional tabular blade at speeds between 100 and 200 rpm under identical hydraulic conditions [16]. The airfoil profile demonstrated a 6% improvement in efficiency over the tabular design, peaking at 140 rpm. In contrast, the present study explored a broader speed range (270–940 rpm) and observed markedly superior efficiency, particularly for the aerodynamic blade, which reached a maximum of 83.5% at 800 rpm. The sharp blade also performed strongly, achieving an optimal efficiency of 79% at 670 rpm, while the round blade attained 78% efficiency. As anticipated, the flat blade lagged behind, with a peak of 76%, supporting its continued use primarily in cost-constrained applications. These results demonstrate that the aerodynamic and sharp blades exhibit strong efficiency gains at elevated speeds, validating their suitability for high-performance CFT applications. The findings also suggest that previous studies may have underestimated the turbine efficiency at higher speeds due to the limited operational range and suboptimal blade-flow interaction models.

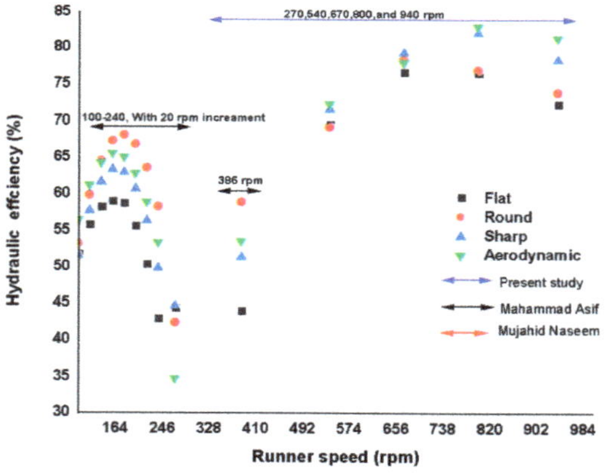

Figure 28. Comparative analysis of blade efficiency trends from prior and current studies.

4. Conclusions

This study has demonstrated that the blade profile geometry exerts a significant influence on the internal flow behavior and hydraulic performance of Cross-Flow Turbines (CFTs). Through high-resolution CFD simulations, it was found that the aerodynamic blade profile, inspired by NACA airfoils, provided superior performance across all runner speeds, achieving a maximum efficiency of 83.5% at 800 rpm. This profile exhibited favorable flow characteristics, including stable velocity fields, low turbulence intensity, and optimal stage-wise energy extraction. The sharp-edged blades also showed high efficiency at intermediate speeds, although with reduced robustness under off-design conditions. Conversely, flat and round blade profiles, while structurally simple and economically favorable, were associated with increased flow detachment, pressure losses, and elevated exit pressures, limiting their performance.

Stage-wise analysis revealed that the first stage dominates energy extraction, contributing up to 70% of the total output for well-optimized profiles. However, the efficiency of the second stage remains critically dependent on the upstream flow quality, which is best preserved by the aerodynamic and round blades. Exit pressure analysis further confirmed that effective pressure recovery and suppression of cavitation-prone regions are directly correlated with blade geometry. The water volume fraction and streamline visualizations supported these conclusions, highlighting the aerodynamic profile's ability to sustain continuous, stable flow with minimal air entrainment or void formation. The findings not only validate the efficiency benefits of aerodynamic profiles in high-speed micro-hydropower applications but also suggest the feasibility of hybrid blade designs that merge structural simplicity with hydrodynamic performance.

Moreover, the optimization of CFT blade designs holds substantial promise for broader deployment within decentralized hybrid renewable energy systems. When integrated with complementary technologies such as solar photovoltaic (PV) and battery storage, these optimized turbines can enhance energy reliability, grid resilience, and cost-effectiveness in off-grid or weak-grid settings. Such systems align strongly with the goals of Sustainable Development Goal 7 (Affordable and Clean Energy), supporting rural electrification, sustainable infrastructure planning, and cross-sector energy policy. Expanding the application of high-efficiency CFTs beyond isolated hydropower contexts into multi-source renewable frameworks presents a critical step toward scalable and inclusive energy solutions.

5. Recommendations

- ☞ Design Optimization: Future turbine designs should prioritize aerodynamic or sharp-edge profiles for improved hydraulic efficiency and internal flow stability, particularly in high-speed operations typical of decentralized energy systems.
- ☞ Hybrid Geometry Exploration: A promising design avenue lies in hybrid blades that integrate a rounded leading edge (to suppress stagnation) with an aerodynamic trailing edge (to minimize separation), balancing manufacturability and performance.
- ☞ Experimental Validation: To further support the simulation results, physical prototype testing under controlled conditions using PIV or LDA techniques is recommended for capturing transient effects and validating the turbulence behavior.
- ☞ Three-Dimensional Modeling: Extending the study to three-dimensional and transient CFD simulations using LES or DES models would help capture secondary flow phenomena and assess unsteady behavior, particularly near blade tips and the runner-shaft interface.
- ☞ Multi-objective Optimization: Incorporate techno-economic metrics into future design frameworks to simultaneously optimize hydraulic performance, material cost, and fabrication feasibility for scalable implementation in remote or resource-constrained regions.

Author Contributions: Both authors contributed to the conceptualization, analysis, and interpretation. E.Y.A. performed intensive analysis, drafting the manuscript and editing, and A.H.T. performed supervision, drafting, reviewing, and editing. All authors have read and agreed to the published version of the manuscript.

Funding: This study received no external funding.

Data Availability Statement: All important data is included in the manuscript.

Acknowledgments: We authors are grateful to NTNU (Norwegian University of Science and Technology) and Institute of Energy (IOE) for their unwavering support for this study.

Conflicts of Interest: The authors declare no conflicts of interest.

Abbreviations

Symbol	Description	Unit
αp	Volume fraction of a particular phase	-
ρ_P	Density of a particular phase	Kg/m^3
μ_P	Viscosity of a particular phase	
S_P	Continuity equation source term for a particular phase	Kg/s
$\vec{V}_r$	relative velocity of the flow	m/s
μ_{eff}	Effective viscosity	kg/m. s
μt	Turbulence viscosity	kg/m. s
μ	Mean viscosity	(kg/m.s)
$\vec{\omega}$	Angular velocity of the flow:	m/s
$\vec{r}$	Radius of the runner	m
$\vec{\alpha}$	Angular acceleration	m^2/s
$\vec{a}$	Linear acceleration	m^2/s
$S_{MP}\nabla$ (-)	Momentum equation source term for a particular phase Del operator	N/m^3
$\dot{P}$	Modified pressure	pa
P	Pressure	Pa
Po	Reference pressure	pa
$\vec{V}$	Absolute velocity of the flow	
T	Transpose	-
t	Time	sec
k	Turbulence kinetic energy	J/kg
ε	Turbulent Dissipation rate	m^3/s^3
ω	Turbulent frequency	s^{-1}
w	Volume fraction of water	-
a	Volume fraction of air	-
C_μ	Constants	-
ρ	Density of the fluid, and	(kg/m^3)
γ_a	Air compressibility coefficient	-
T	Torque	N.m
τ	The total stress tensor	Pa
s	Surface area	m^2
n	Unit Vector Perpendicular to the Surface	
r	Radial position of the vector	m
a	Unit vector parallel to the rotation axis	
$P_{turbine}$	Shaft power	kN
P_{water}	Water power	kN
H	Effective net head	m
Q	Flow rate	m^3/s
γ	Specific weight	N/m^3
η	Hydraulic efficiency	%
V	Absolute water velocity	m/s
U	Peripheral blade velocity	m/s
W	Relative blade velocity	m/s
V_f	The radial velocity component	m/s
Vu	The tangential velocity component	m/s
α	Angle of attack: 0°	deg
β	Blade angle	deg
θ	Angular (azimuthal)position	deg
h	Characteristic grid spacing	mm
r_{ij}	Refinement ratio	-
N	Number of cells	-

ε_{ij}	Relative error	%
q_{n+1}	Observed order of accuracy	-
τ_{exact}	Richardson's extrapolated torque	

Abbreviations	**Description**
PIV	Particle Image Velocimetry
LDA	Laser Doppler Anemometry
CFT	Cross-flow turbine
MHP	Micro hydropower
CFD	Computational fluid dynamics
SST	Shear-stress turbulence
WVF	Water volume fraction
GCI	Grid convergence index
FS	Safety Factor

References

1. International Energy Agency (IEA). Access to Electricity. Paris, France. Available online: http://www.iea.org/weo/electricity.asp (accessed on 24 January 2024).
2. The High-Level Political Forum Goals Energy Action. Available online: https://hlpf.un.org/ (accessed on 19 February 2025).
3. Eludoyin, E.O. The Nature of Sustainable Energy Access Transitions: Realities and possibilities for Lagos, Nigeria. Ph.D. Thesis, UCL Energy Institute, University College London, London, UK, 2018; 421p. Available online: http://discovery.ucl.ac.uk/10057177/7/Eludoyin_10057177_thesis_redacted_photo.pdf (accessed on 13 July 2023).
4. Kaunda, C. Energy situation, potential and application status of small scale hydropower systems in Malawi. *Renew. Sustain. Energy Rev.* **2013**, *26*, 1–19. [CrossRef]
5. Del Rio, J.S.; Galvis-Holguin, S.; Hincapié-Zuluaga, D.; Chica, E. Effect of hydrodynamically designed blades on the efficiency of a michell-banki turbine. *Int. J. Renew. Energy Res.* **2020**, *10*, 1165–1173. [CrossRef]
6. Sammartano, V.; Aricò, C.; Carravetta, A.; Fecarotta, O.; Tucciarelli, T. Bank-Michell optimal design by computational fluid dynamic testing and hydynamic analysis. *Energies* **2013**, *6*, 2362–2385. [CrossRef]
7. Muchtar, M.; Rasyid, S.; Nur, R. Welding fixture design for the production of cross-flow turbine runner. *AIP Conf. Proc.* **2018**, *1977*, 030020. [CrossRef]
8. Adhikari, R.; Wood, D. The design of high efficiency crossflow hydro turbines: A review and extension. *Energies* **2018**, *11*, 267. [CrossRef]
9. Baidar, B.; Koirala, R.; Neopane, H.P.; Shrestha, M.V.; Thapa, B. Strategic rehabilitation of the earthquake affected microhydropower plants in Nepal. *IOP Conf. Ser. Earth Environ. Sci.* **2016**, *49*, 102003. [CrossRef]
10. Yildiz, V. *Numerical Simulation Model of Run of River Hydropower Plants: Concepts, Numerical Modeling, Turbine System and Selection, and Design Optimization*; University of California: Irvine, CA, USA, 2015; p. 84.
11. Assefa, E.Y.; Tesfay, A.H. Effect of Blade Number on Internal Flow and Performance Characteristics in Low-Head Cross-Flow Turbines. *Energies* **2025**, *18*, 318. [CrossRef]
12. Yadav, V.K.; Singal, S.K. Performance analysis of cross-flow turbine: Variation in shaft diameter. In *Development of Water Resources in India. Water Science and Technology Library*; Springer: Cham, Switzerland, 2017; Volume 75, pp. 487–497.
13. Guo, Q.; Zhou, L.; Wang, Z. Numerical evaluation of the clearance geometries effect on the flow field and performance of a hydrofoil. *Renew. Energy* **2016**, *99*, 390–397. [CrossRef]
14. Adanta, D.; Sari, D.P.; Syofii, I.; Prakoso, A.P.; Saputra, M.A.A.; Thamrin, I. Performance Comparison of Crossflow Turbine Configuration Upper Blade Convex and Curvature by Computational Method. *Civ. Eng. J.* **2023**, *9*, 154–165. [CrossRef]
15. Desai, V.R.; Aziz, N.M. An experimental investigation of cross-flow turbine efficiency. *J. Fluids Eng. Trans. ASME* **1994**, *116*, 545–550. [CrossRef]
16. Anand, R.S.; Jawahar, C.P.; Bellos, E.; Malmquist, A. A comprehensive review on Crossflow turbine for hydropower applications. *Ocean. Eng.* **2021**, *240*, 110015. [CrossRef]
17. Woldemariam, E.T.; Lemu, H.G.; Wang, G.G. CFD-driven valve shape optimization for performance improvement of a micro cross-flow turbine. *Energies* **2018**, *11*, 248. [CrossRef]
18. Rasmussen, I.J. Design of a Francis Model Test Rig at Kathmandu University. Master's Thesis, Institutt for Energi-og Prosessteknikk, Trondheim, Norway, 2014.
19. Sinagra, M.; Picone, C.; Aricò, C.; Pantano, A.; Tucciarelli, T.; Hannachi, M.; Driss, Z. Impeller optimization in crossflow hydraulic turbines. *Water* **2021**, *13*, 313. [CrossRef]

20. Chelabi, M.A.; Dobrotvorskiy, S.; Basova, Y.; Aleksenko, B.A.; Edl, M.; Zdebor, J.; Machado, J. Influence of the Main Geometrical Parameters on the Design and Performance of Mixed Inflow Turbines. *Appl. Sci.* **2022**, *12*, 12165. [CrossRef]
21. Naseem, M.; Saleem, A.; Naseem, M.S. Investigation of blade design parameters for performance improvement of hydraulic cross flow turbine. *Ocean Eng.* **2022**, *257*, 111663. [CrossRef]
22. Urame, C.; Hoole, P.R. Design and Implementation of Hybrid Pico-Hydro—Photovoltaic (PV) Solar Power Plant in Massy-Gahuku LLG. *Eur. J. Electr. Eng.* **2020**, *22*, 395–403. [CrossRef]
23. Sedlar, M.; Ji, B.; Kratky, T.; Rebok, T.; Huzlik, R. Numerical and experimental investigation of three-dimensional cavitating flow around the straight NACA2412 hydrofoil. *Ocean Eng.* **2016**, *123*, 357–382. [CrossRef]
24. Georgescu, A.M.; Georgescu, S.C.; Degeratu, M. Experimental setup designed for testing a cross-flow water turbine in a wind tunnel. *UPB Sci. Bull. Ser. D Mech. Eng.* **2008**, *70*, 191–200.
25. Japan International Cooperation Agency. Guideline and Manual for Hydropower Development Vol. 2. Small Scale Hydropower. 2011. Available online: https://openjicareport.jica.go.jp/pdf/12024881_01.pdf (accessed on 3 April 2025).
26. Oberkampf, W.L.; Trucano, T.G. Verification and validation in computational fluid dynamics. *Prog. Aerosp. Sci.* **2002**, *38*, 209–272. [CrossRef]
27. Menter, F.R.; Kuntz, M.; Langtry, R. Ten Years of Industrial Experience with the SST Turbulence Model Turbulence heat and mass transfer. *Turbul. Heat Mass Transf.* **2004**, *4*, 625–632.
28. Phillips, T.S.; Roy, C.J. Richardson extrapolation-based discretization uncertaint estimation for computational fluid dynamics. *J. Fluids Eng. Trans. ASME* **2014**, *136*, 121401. [CrossRef]
29. Ahn, S.Y.; Kim, K.Y. Thermal performance of t-shaped obstacles in a solar air heater. *Processes* **2020**, *8*, 1305. [CrossRef]
30. Kaunda, C.S.; Kimambo, C.Z.; Nielsen, T.K. A numerical investigation of flow profile and performance of a low cost crossflow turbine. *Int. J. Energy Envrion.* **2014**, *5*, 275–296.
31. Javaherchi, T.; Stelzenmuller, N.; Seydel, J.; Aliseda, A. Experimental and Numerical Analysis of a Scale-Model Horizontal Axis Hydrokinetic Turbine. In Proceedings of the 2nd Marine Energy Technology Symposium METS2014, Seattle, WA, USA, 15–18 April 2014.
32. Menter, F.R.; Sechner, R.; Matyushenko, A. Best Practice: RANS Turbulence Modeling in ANSYS CFD. 2021. 295p. Available online: https://www.ansys.com/resource-center/technical-paper/best-practice-rans-turbulence-modeling-in-ansys-cfd (accessed on 19 February 2025).
33. Ibrahim, B.; Chattha, J.A.; Asif, M. Optimization of Blade Profiles of Cross Flow Turbine (CFT). *Int. J. Power Energy Convers. IJPEC* **2018**, *9*, 311–326. [CrossRef]

energies

MDPI

Article

Fluid Dynamics Analysis of Flow Characteristics in the Clearance of Hydraulic Turbine Seal Rings

Leilei Chen [1,2], Wenhao Wu [3], Jian Deng [1,2], Bing Xue [1,2], Liuming Xu [1,2], Baosheng Xie [3] and Yuchuan Wang [3,*]

1 Hubei Technology Innovation Center for Smart Hydropower, Wuhan 430000, China; chen_leilei1@ctg.com.cn (L.C.); deng_jian@cypc.com.cn (J.D.); xue_bing@ctg.com.cn (B.X.); xu_liuming@ctg.com.cn (L.X.)

2 China Yangtze Power Co., Ltd., Yichang 443002, China

3 College of Water Conservancy and Construction Engineering, Northwest A&F University, Yangling 712100, China; wwh000@nwafu.edu.cn (W.W.); x1b1s103@nwafu.edu.cn (B.X.)

* Correspondence: wyc@nwafu.edu.cn

Abstract

The hydraulic turbine serves as the cornerstone of hydropower generation systems, with the sealing system's performance critically influencing energy conversion efficiency and operational cost-effectiveness. The sealing ring is a pivotal component, which mitigates leakage and energy loss by regulating flow within the narrow gap between itself and the frame. This study investigates the intricate flow dynamics within the gap between the sealing ring and the upper frame of a super-large-scale Francis turbine, with a specific focus on the rotating wall's impact on the flow field. Employing theoretical modeling and three-dimensional transient computational fluid dynamics (CFD) simulations grounded in real turbine design parameters, the research reveals that the rotating wall significantly alters shear flow and vortex formation within the gap. Tangential velocity exhibits a nonlinear profile, accompanied by heightened turbulence intensity near the wall. The short flow channel height markedly shapes flow evolution, driving the axial velocity profile away from a conventional parabolic pattern. Further analysis of rotation-induced vortices and flow instabilities, supported by turbulence kinetic energy monitoring and spectral analysis, reveals the periodic nature of vortex shedding and pressure fluctuations. These findings elucidate the internal flow mechanisms of the sealing ring, offering a theoretical framework for analyzing flow in microscale gaps. Moreover, the resulting flow field data establishes a robust foundation for future studies on upper crown gap flow stability and sealing ring dynamics.

Keywords: turbine; seal ring; shear flow; flow analysis; micro-gap flow; CFD simulation

1. Introduction

As the core power-conversion device in modern hydropower systems, the investigation of the internal fluid characteristics of hydraulic turbines has been a long-standing research topic [1]. A turbine's fluid-dynamic performance directly determines both its energy-conversion efficiency and the levelized cost of electricity over the plant's entire life cycle. According to the IEA's World Energy Outlook 2022, hydropower accounts for 61.8% of global renewable-generation capacity [2], a leading position derived from its superior conversion efficiency, operational flexibility, and large-scale storage capability.

However, as individual unit capacities exceed the megawatt scale and design heads extend to around 800 m, transient hydraulic excitations within the system have increasingly manifested, making the safe and stable operation of the units paramount.

Seal rings are critical sealing components in rotating machinery: the narrow annular gap between the ring and the upper bracket plays an indispensable role in maintaining turbine efficiency and stability. Research has shown that even a small gap at this location can lead to efficiency loss, and the degree of loss increases with the gap size. The loss is particularly severe when there are sediments in the water [3]. The presence of the sealing ring significantly affects pressure fluctuations in the gap flow and alters the axial hydraulic thrust during the hydraulic transient process [4].

In super-large-scale Francis turbines, although the clearance is very small and the fluid volume in these regions is several orders of magnitude smaller compared to other dimensions in the hydraulic turbine, the flow characteristics in these areas can have a significant impact on the overall flow within the machine. Studies have found that the leakage flow in the sidewall gap contributes significantly to the fluid-induced rotor dynamics forces. Swiss scholar Peter Dorfler et al. [5] highlighted the effects of gap flow in the sealing ring, noting that the gap flow phenomenon has an important interference effect on the rotational motion between the rotor and the stationary components, thereby influencing the pressure distribution in the sealing gap. The resulting hydraulic reaction forces form a feedback mechanism on the rotor. This transient hydraulic excitation phenomenon is particularly pronounced under variable-speed operating conditions, where nonlinear flow effects within micro-gaps of 0.5–5 mm can amplify pressure pulsations to two to three times those of steady-state conditions [6]. Such transient hydraulic excitations are particularly pronounced under variable-speed operation.

The flow through the seal-ring gap is characterized by strong multi-physics coupling: first, seal rings are interference-fitted via a thermal expansion process, imparting a preload; second, the combined effects of cavitation and abrasion in sediment-laden water shorten sealing life; and third, the rotational effects of the runner crown introduce positional sensitivity in pressure pulsations, further increasing the risk during normal operation.

The flow within the sidewall gap is highly complex and can exhibit various states and structures. The labyrinth seal consists of a series of narrow annular gaps and chambers between the rotating and stationary components. The flow within these regions is typically turbulent. Flow characteristics in this region not only govern leakage control but also directly influence the uniformity of pressure distribution and the degree of energy loss. Existing studies indicate that in rotating gap cavities, high-speed fluid motion induces significant hydrodynamic pressure effects, thereby enhancing film lift and stiffness. For example, in the downstream cavity of a supercritical CO_2 compressor impeller and in seal-ring chambers, researchers have observed the formation of enlarged high-vorticity regions as rotational speed and pressure ratio increase—an outcome of intensified flow instability. The expansion of these high-vorticity zones demonstrates that high-speed rotation in small gaps generates strong vortex structures and markedly increases the friction-resistance coefficient [7]. Overall, rotating small-gap flows are typically accompanied by complex tangential convergence and vortex structures, with flow-field features that differ markedly from those under stationary conditions.

In rotating shear flows, boundary-layer instabilities are key to flow evolution. Cui et al. [8] point out in their review that in thin oil films between multiple rotating disks, the interplay of squeeze and shear effects renders the laminar-to-turbulent transition exceedingly complex. Du et al. [9] further demonstrate that in rotating-stationary disk cavities, turbulence tends to first emerge in the stationary disk's boundary layer. As the Reynolds number rises, spiral-instability modes amplify and interact with axisymmetric wave modes, leading to localized turbulence. These findings highlight that three-dimensional vortex structures—such as spiral waves—dominate the unsteady evolution of rotating cavity flows and are crucial to understanding turbulence onset and instability mechanisms.

Numerous researchers and academic authorities have formulated theoretical models and solved analytical equations for the flow in such regions. We consider the coaxial cylindrical gap flow with the inner cylinder rotating and the outer cylinder fixed, while superimposing axial flow driven by an axial pressure gradient. This basic flow is steady and laminar and can be decomposed into a radial no-flow, circumferential Couette component, and an axial Poiseuille component. Starting from the incompressible Navier–Stokes equations and assuming axial and circumferential translation invariance, the benchmark equation for no radial flow can be solved to obtain the laminar velocity analytical expression. Martinand et al. [10] provided the benchmark flow analytical expression for the ratio of the inner and outer cylinder radii. In the case of a narrow gap, the Taylor–Couette system can be approximated as a parallel plate model. Recently, Nagata et al. [11] performed a theoretical analysis of this limit by applying Cartesian coordinate processing in the slit limit. A linear stability analysis of the region was also conducted based on the benchmark laminar flow. It was found that the axial flow has a stabilizing effect on fluid stability: stronger axial flow suppresses the growth of radial-circumferential vortices, thus increasing the critical Taylor number (Ta). Even weak axial flow can sometimes alter the mode shape, resulting in spiral or wave-type instability modes [12].

In recent years, CFD simulation has become an essential tool for studying micro-gap flows. Capturing vortex structures and unsteady features in small rotating gaps requires high-resolution unsteady simulations [13,14]. Moore [15], through three-dimensional CFD rotor-dynamic analysis, found that rotation not only drives tangential flow but also induces complex vortex structures that directly affect leakage rates and sealing forces; simulations show helical rising trajectories of flow within the seal, and vortex intensity increases with speed. Ding et al. [16] used CFD to analyze the sealing performance of floating-ring seals with and without helical grooves, comparing their flow-field differences. However, most studies still focus on macroscopic leakage rates and pressure drops, with limited attention to microscopic pressure distributions inside dynamic-pressure grooves. Jürgen Schiffer et al. [17] evaluated the static pressure in the sidewall gap of the rotor along various coordinate points uniformly distributed across the labyrinth seal flow passage. They suggested that the resulting pressure distribution could serve as the basis for calculating axial thrust. In the micro-gaps between the rotor and stator, rotating-wall effects can have a significant impact on leakage-flow characteristics. Kim et al. [18] compared LES and RANS for straight-through and stepped labyrinth seals, finding that LES accurately captured a 7% reduction in leakage coefficient due to rotation, while RANS underestimated this effect. Notably, LES revealed relaminarization trends in the rotating gap's low-Re region that RANS overpredicted as turbulent kinetic energy. This suggests that, for small-gap flows—especially at low Reynolds numbers—traditional RANS models may mischaracterize flow features and higher-fidelity models are required to reflect rotating-wall mitigation and turbulence-decay effects. For the engineering numerical simulation of small-gap Taylor–Couette flow, although high-fidelity DNS/LES can provide the most comprehensive flow field information, the SST k-ω model has become the most widely used and stable URANS turbulence model in industrial applications due to computational resource limitations. It balances boundary layer capture accuracy and numerical stability and has achieved good results in multiple comparative experiments and simulation studies [19,20]. CFD studies of multi-cavity labyrinth seals (including rotor-seal systems) have further elucidated how rotation-induced flows and vortices influence sealing performance. Jia et al. [21] investigated the transient flow in variable-speed rotor–labyrinth systems, showing that rotor vibration increases leakage, although this effect diminishes at high speeds, and that labyrinth-generated aerodynamic forces enhance rotor stability, with coupling effects weakening as speed increases. Zhang et al. [22] conducted a detailed simulation of a hole diaphragm labyrinth seal (HDLS) at different eccentric frequencies and rotational speeds, finding that the orifice structure introduces additional turbulence-dissipation sources in each cavity—increasing turbulent kinetic energy—and observing backflow leakage at high eccentric frequencies, establishing an exponential relationship between leakage rate and speed. These studies reveal the nonlinear phenomena of rotation-induced vortices and flow fields in multi-cavity seals, offering new perspectives for predicting sealing performance and leakage.

Rotationally induced vortex structures—such as spiral and saddle vortices—are crucial to flow characteristics. In rotating channels, Coriolis forces excite counter-rotating vortex pairs, leading to significantly different flow and heat-transfer performance between leading and trailing edges [23]. Xu et al. [24] found that in high-speed rotating cylindrical gas-film seals, converging gaps and dynamic-pressure grooves generate strong hydrodynamic effects that greatly increase film lift and stiffness. Moreover, unsteady simulations by Perini et al. [25] show that "hot-spot" vortices orbiting the rotor can be observed in turbine-runner seal cavities—large-scale structures unrelated to blade excitation and rotating at near-rotor speed. These studies demonstrate that common vortical structures in rotating micro-gap systems dominate unsteady behavior and energy transfer, directly affecting leakage flow and vibration characteristics.

In summary, most of the literature focuses on gas-film seals or simple disk-cavity flows and employs steady or quasi-steady models that often neglect strong unsteady instabilities—such as vortex shedding and flow pulsations—that may occur in the gaps. The expansion of high-vorticity zones due to flow-instability enhancement suggests the potential for self-excited oscillations in rotating systems. High-fidelity studies on flow and stability in rotating micro-gaps under water-environment conditions are still scarce—particularly regarding the coupled, multi-scale effects introduced by short channel heights and high rotational speeds under complex boundary conditions.

This study is based on the actual problem of sealing ring failure during operation. The phenomenon is essentially caused by the flow characteristics at this location (high shear, high pressure pulsations, vortices, and turbulence excitation). Understanding the flow mechanism in this gap is the foundation and starting point of all subsequent work. The unsteady characteristics of this region (such as pressure pulsations, axial water thrust variations, vortex-induced vibrations, etc.) can persist under non-extreme operating conditions, particularly in typical operating conditions such as startup, shutdown, and partial load, where they are more pronounced. The flow characteristics within the micro-gap determine the pressure pulsations of the sealing ring and the distribution of fluid–structure coupling loads, which form the basis for subsequent dynamic analysis, vibration prediction, and fatigue life assessment. Building on this state of the art, the present study will focus on simulating and analyzing micro-gap flows in hydroturbine seal rings. Unlike prior work focusing primarily on gas-film seals, we will establish a numerical model in ANSYS Fluent that incorporates rotating walls and water-pressure environments. Through detailed CFD simulations, we will characterize vortex-structure distributions, pressure pulsations, and unsteady flow evolution in the micro-gap—providing scientific insight into actual seal-ring flows, offering substantive guidance for safe and stable turbine operation and yielding theoretical and practical references for flow analysis and the engineering design of other rotating machinery.

2. Theoretical Model Analysis

2.1. Flow Model Establishment

A theoretical analysis was conducted on the physical model of the turbine used in this study. The geometric parameters of the fluid domain in the seal-ring gap are as follows: inner wall diameter 9315 mm, gap width h = 4 mm, outer wall diameter 9323 mm, inner wall rotational speed—the runner speed—76 rpm, and channel height L = 0.26 m. The combination of an extremely small gap width ($h/R_1 \approx 0.00086$, the gap can be approximated using a slit treatment) and a short channel height (L/h = 65), together with high-speed rotation (inner wall tangential velocity ≈ 37.06 m/s), yields flow behaviors with pronounced non-classical characteristics. Classical theories of narrow-gap flow are inadequate to fully describe the flow regimes under these extreme geometric and dynamic conditions. Therefore, this study employs targeted analytical methods to systematically derive fundamental flow characteristics, such as velocity and pressure distributions, while placing special emphasis on the complex, multiscale coupling effects introduced by the short channel height and high rotation speed. A critical stability analysis of rotation-induced phenomena is also presented to provide theoretical support for the safe, stable, and efficient operation of turbine seals. Figure 1 is a cross-sectional diagram of the assembly of the sealing ring component.

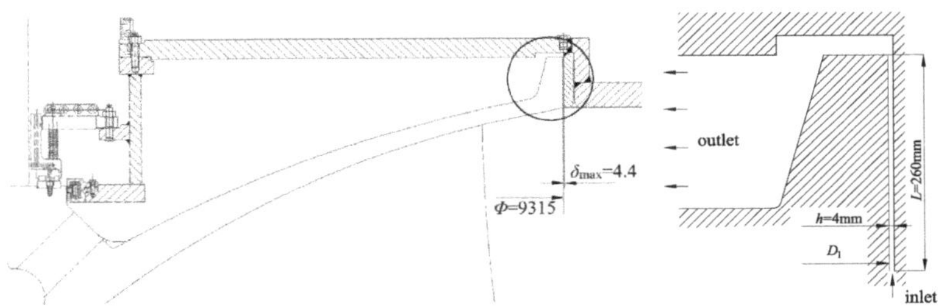

Figure 1. Seal-ring clearance structure diagram.

Assuming the fluid is an incompressible Newtonian fluid with a density of ρ = 1000 kg/m^3 and viscosity of μ = 0.001 Pa·s and that the flow is steady and laminar, a Cartesian coordinate system was adopted, with the z-axis aligned with the channel height. The inner wall rotates at ω = 7.96 rad/s, while the outer wall remains stationary; the fluid is driven upward along the axial (z) direction by a pressure gradient dp/dz and gravity. Because the gap width $h \ll R_1$, the annular seal-ring gap may be approximated as planar flow. However, the short channel height L = 0.26 m combined with high-speed rotation introduces unique multi-scale effects: the radial scale is far smaller than the axial and circumferential scales, and the tangential shear time ($h/(\omega R_1) \approx 0.1$ ms) is orders of magnitude faster than the axial flow evolution time on the order of seconds ($L/u_{z,\text{avg}} \approx 0.1$–1 s). These cross-scale geometric and dynamic conditions produce a complex coupling of flow behaviors controlled by different physical mechanisms, requiring special analytical treatment to capture their underlying physics. Figure 2 is the theoretical model established based on the actual physical structure.

2.2. Flow Characterization

2.2.1. Tangential Velocity Distribution

The annular slit can be approximated as a planar Couette flow due to the gap width $h = R_2 - R_1 << R_1$. Define the local coordinates $y = r - R_1$, where $0 \le y \le h$. Under the rotational drive of the inner wall, the tangential velocity u_θ (y) satisfies the following boundary conditions:

$$\begin{cases} y = 0 : u_\theta = \omega R_1 \approx 37.06\ \mathrm{m/s} \\ y = h : u_\theta = 0 \end{cases} \tag{1}$$

according to the Navier–Stokes equations in cylindrical coordinates.

$$\rho\left(u_r\frac{\partial u_\theta}{\partial r} + \frac{u_\theta}{r}\frac{\partial u_\theta}{\partial \phi} + \frac{u_r u_\theta}{r}\right) = -\frac{1}{r}\frac{\partial p}{\partial \theta} + \mu\left[\frac{\partial}{\partial r}\left(\frac{1}{r}\frac{\partial (r u_\theta)}{\partial r}\right) + \frac{1}{r^2}\frac{\partial^2 u_\theta}{\partial \theta^2} + \frac{\partial^2 u_\theta}{\partial z^2} + \frac{2}{r^2}\frac{\partial u_r}{\partial \theta}\right] \tag{2}$$

The simplified control equation is

$$\frac{\partial}{\partial r}\left(\frac{1}{r}\frac{\partial (r u_\theta)}{\partial r}\right) = 0 \tag{3}$$

Substituting the boundary conditions, the solution is obtained using

$$u_\theta(r) = \frac{\omega_1 R_1 (R_2 - r) + \omega_2 R_2 (r - R_1)}{R_2 - R_1}. \tag{4}$$

Because the acceleration of the outer cylinder is zero, $\omega_2 = 0$, the final result is

$$u_\theta(y) = \omega R_1 \left(1 - \frac{y}{h}\right). \tag{5}$$

Expanding the above solution under the slit conditions, extracting the classical linear Couette component along with the nonlinear correction introduced by the geometric curvature [26], the result is

$$u_\theta(y) \approx \omega \cdot y \left[1 + \frac{y}{R_1} - \frac{h}{2R_1} + \mathcal{O}\left(\left(\frac{y}{R_1}\right)^2\right)\right]. \tag{6}$$

The four terms in parentheses represent the simple planar Couette flow linear term, the curvature effect from the annular geometry, the constant correction term, and the higher-order geometric nonlinear term.

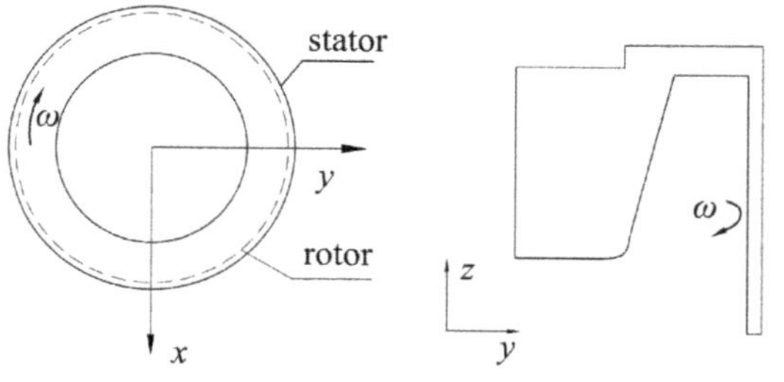

Figure 2. Theoretical calculation model. The arrow indicates the direction of rotation of the rotor, and the angular velocity of rotation is ω.

After considering the narrow gap, the tangential velocity distribution becomes nonlinear, and the shear rate can reach up to $du_\theta/dy \approx -9250\ \mathrm{s}^{-1}$. The coupling of the ultra-high shear rate within the radial scale (4 mm) with the high rotation speed in the annular scale (4.66 m) significantly enhances the friction loss, while the rapid millisecond variation in the tangential shear time further highlights the dynamic non-uniformity. In addition, the high-speed rotation of the inner wall may trigger instabilities similar to the Taylor–Couette flow, and its effect on flow stability needs to be further analyzed.

2.2.2. Axial Velocity Distribution

The axial speed $u_z(y)$ is driven by a pressure gradient and the control equation is

$$\mu \frac{d^2 u_z}{dy^2} = \frac{dp}{dz} \tag{7}$$

Boundary conditions:

$$\begin{cases} y = 0, u_z = 0 \\ y = h, u_z = 0 \end{cases} \tag{8}$$

Solution:

$$u_z(y) = \frac{1}{2\mu} \frac{dp}{dz} y(y - h) \tag{9}$$

The velocity profile is parabolic with a maximum value:

$$u_{z,\max} = -\frac{h^2}{8\mu} \frac{dp}{dz} \tag{10}$$

The short runner height L = 0.26 m may limit the full development of the axial flow. For this reason, the inlet length L_e is introduced, i.e., the axial distance required for the velocity profile to develop from the initial state (usually uniformly distributed) to the fully developed state (parabolic distribution) after the fluid enters the gap from the inlet. The inlet length is estimated using the following empirical equation:

$$L_e \approx 0.05\text{Re} \cdot h,\ \text{Re} = \frac{\rho u_{z,\text{avg}} h}{\mu}. \tag{11}$$

Assuming an axial average velocity of $u_{z,\text{avg}} \approx$ 0.1–1 s, then Re $\approx$ 400–4000, thus

$$L_e \approx 0.05 \times (400 \sim 4000) \times 0.004 = 0.08 \sim 0.8\ \text{m}. \tag{12}$$

The runner height L = 0.26 m may be less than or close to L_e, indicating that the flow may be underdeveloped, the axial flow evolution time is limited by the short runner constraints, and the velocity profiles deviate from the theoretical parabolic distribution, which affects the prediction of the flow rate and pressure loss.

2.2.3. Pressure Distribution

(1) Axial pressure

The pressure in the z-direction along the flow $p(z)$ is a combination of inlet pressure, flow pressure drop, and gravity pressure drop:

$$p(z) = p_0 + \left(\frac{\partial p}{\partial_{\text{zflow}}} - \rho g \right) z \tag{13}$$

where gravity pressure drops,

$$\Delta p_g = \rho g L \approx 2550\text{Pa}. \tag{14}$$

Although Δp_g is much smaller than the flow pressure drop (typically on the order of 10^4–10^5 Pa), short flow paths highly concentrate pressure gradient changes, making axial pressure distributions more sensitive and potentially amplifying the competing effects of gravity and flow pressure drop locally (e.g., at the inlet or outlet).

(2) Radial pressure

This part of the pressure distribution is mainly induced by centrifugal force:

$$\frac{\partial p}{\partial r} = \frac{\rho u_\theta^2}{r} \approx \rho \omega^2 R_1 \tag{15}$$

Radial differential pressure:

$$\Delta p_r \approx \rho \omega^2 R_1 h \approx 1180\text{Pa} \tag{16}$$

The pressure gradient along the radial direction is not very pronounced because the radial pressure difference differs from the axial pressure difference by at least two orders of magnitude.

2.3. Flow Stability Analysis at Multiple Scales

To more comprehensively describe the flow characteristics of the gap channel, we introduced the calculation of the Reynolds number (*Re*) to determine whether the fluid is in a laminar or turbulent state [27]. The Reynolds number is a dimensionless parameter, defined as

$$Re = \frac{\rho v D}{\mu}. \tag{17}$$

In the annular gap channel, due to the combined effects of the rotating inner wall and the axial pressure gradient, the fluid possesses both tangential and axial velocity components. Therefore, we calculated the tangential Reynolds number and axial Reynolds number separately and conducted a comprehensive analysis of the overall flow state:

$$\begin{array}{l} Re_\theta = \frac{\rho u_{\theta,\max} h}{\mu} = \frac{1000 \times 37.06 \times 0.004}{0.001} = 148,240 \\ Re_z = \frac{\rho u_{z,\text{avg}} D_h}{\mu} = \frac{1000 \times 1 \times 0.004}{0.001} = 4000 \end{array}. \tag{18}$$

It can be observed that both the axial and tangential Reynolds numbers are greater than the critical value of 2300 for laminar flow, and the axial Reynolds number is even higher in actual flow. Therefore, it can be concluded that the flow is in a turbulent state.

As discussed in Section 2.2, the high-speed rotation of the inner wall may induce an instability similar to the Taylor–Couette flow, which is defined as the fluid motion between two concentric cylinders, where centrifugal forces may induce an instability when the inner cylinders rotate, forming periodic vortices (Taylor vortices). To quantify this instability, the Taylor number *Ta* is introduced and defined as

$$Ta = \frac{\omega^2 R_1 h^3}{\nu^2} \tag{19}$$

where $\nu = \frac{\mu}{\rho} = 10^{-6}\ \text{m}^2/\text{s}$. Substitute parameters:

$$Ta = \frac{(7.96)^2 \times 4.6575 \times (0.004)^3}{(10^{-6})^2} \approx 1.2 \times 10^{10} \tag{20}$$

For the slit ($h << R_1$), the critical Taylor number is $Ta_c \approx 1700$ [28,29]. Due to $Ta >> Ta_c$, the flow is highly unsteady and may transition from laminar to turbulent flow, forming Taylor vortices. However, the short runner height $L = 0.26$ m ($L/h = 65$) is much smaller than the circumferential vortex wavelength ($2\pi R_1 \approx 29.3$ m) and may inhibit the axial development of the vortex. The constraining effect of finite-length flow channels on Taylor–Couette flow has been explored in existing studies. Based on this, a modified Taylor number is proposed to introduce a short flow channel constraint factor:

$$Ta_{\text{corrected}} = Ta \cdot \left(\frac{L}{h}\right)^{-1} \approx 1.8 \times 10^{8}. \tag{21}$$

The correction still exceeds the critical value, indicating that the flow is in the transition state between laminar and turbulent. Multi-scale effects are particularly prominent here: the rotational dynamics induces instability at a radial scale of 4 mm and an annular scale of 4.66 m, while the short flow channel at an axial scale of 0.26 m inhibits the development of vortices. The difference between the tangential shear time in milliseconds and the axial evolution time in seconds further complicates the flow behaviour. This analysis reveals the modulation mechanism of rotational stability by short flow channels, providing a novel perspective for slit flow studies.

To validate the applicability of the critical Taylor number derived from engineering experience and to clarify the type of unstable mode under the current operating conditions, a linear perturbation method can be further employed to construct an eigenvalue problem. The spectral method was used to solve for the system's instability characteristic frequency and growth rate under the slit approximation, thus providing a theoretical explanation for the critical Taylor number and the vortex structures observed in practice.

The total velocity field is decomposed into the base flow and small perturbations:

$$u(y,\theta,z,t) = \mathbf{U}(y) + \varepsilon u'(y,\theta,z,t). \tag{22}$$

Here, $\mathbf{U}(y)$ is the base flow velocity vector, which includes the azimuthal Couette component and the axial Poiseuille component; $\varepsilon << 1$ is the perturbation amplitude coefficient; u' (y, θ, z, t) is the velocity perturbation component. It is assumed that the disturbance has a modal structure with axial wavenumber k and azimuthal mode number n, and its spatiotemporal evolution is described by a complex exponential term:

$$u'(y,\theta,z,t) = \hat{u}(y) \cdot \exp[i(kz + n\theta - \omega t)]. \tag{23}$$

Here, $\hat{u}(y)$ is the amplitude distribution function of the disturbance in the gap direction y; k is the axial wavenumber, describing the spatial period of the disturbance in the z direction; n is the azimuthal mode number, where $n = 0$ represents an axisymmetric Taylor vortex, and $n \neq 0$ represents a helical vortex; ω is the complex frequency, with the real part representing the oscillation frequency and the imaginary part reflecting the growth rate ($\text{Im}\,\omega > 0$) or decay rate ($\text{Im}\,\omega < 0$) of the disturbance.

By substituting the disturbance forms of the above two equations into the incompressible Navier–Stokes equations, we obtain

$$\mathcal{L}[\hat{u}(y)] = i\omega\hat{u}(y). \tag{24}$$

The solution of this eigenvalue problem includes the critical Taylor number *Tac*, the critical Reynolds number *Rec*, and the frequency and growth rate of the corresponding unstable modes, which are used to determine the instability threshold of the base flow and the type of vortex.

3. Numerical Simulation

3.1. Computational Model

Computational fluid dynamics (CFD) has become an indispensable tool for analyzing complex flow phenomena in hydraulic machinery, especially in regions where experimental measurements are difficult to perform. In this study, a three-dimensional CFD simulation of the flow within the narrow annular gap between the seal rings and the upper bracket was carried out using ANSYS Fluent 2022R1 version. Owing to the significant shear and rotational effects involved, this section aims to validate theoretical predictions, reveal detailed flow structures, and further analyze the gap flow characteristics of the seal rings from a visualization perspective.

To simplify the computational model, the annular fluid domain is partitioned in recognition of its periodic symmetry: the full 360° geometry is divided into 36 equal 10° sectors, and one such 1/36 sector is selected as the computational domain. Periodic boundary conditions are imposed on its two radial-cut faces to emulate the full ring. A structured mesh is employed, with refinement near all walls to capture boundary-layer effects and high shear rates. This study focuses on the flow development from the gap inlet up to the point where the flow enters the cover cavity.

In Fluent Meshing, the Poly-Hexcore hybrid meshing method was used to mesh the computational model. Structured hexahedral meshes were used in the internal flow field or regular regions, while polyhedral mesh elements were applied in complex geometric surfaces or irregular regions. Ten layers of boundary layer were added near the wall to capture boundary effects and high shear rates. A local refinement method was used to refine the inlet section. Data monitoring was performed at three measurement points along the radial distribution of the middle height of the inlet section for three sets of meshes, verifying mesh independence and ensuring the convergence and accuracy of the data (see Table 1).

Table 1. Discretization error for numerical study.

Measurement Parameters	Coarse	Medium	Fine	e_{21} (%)	GCI_{21} (%)
Y-Velocity at Point 1 (m/s)	10.2148	9.38755	9.03785	3.869	3.54
Y-Velocity at Point 2 (m/s)	2.70661	3.06686	3.08544	0.602	0.041
Y-Velocity at Point 3 (m/s)	0.286973	0.385668	0.390028	1.118	0.065
Mass Flow Rate at inlet (kg/s)	54.73	53.97	53.58	0.728	0.96
Mass Flow Rate at outlet (kg/s)	54.73	53.97	53.58	0.728	0.96

The total number of mesh cells for the three sets is 5,150,000, 12,906,281, and 41,181,751, with corresponding $y+$ values of 29, 16, and 0.7. It can be observed that the results of the computational model with medium mesh density stabilized. Considering the computational resources, the use of fine mesh is acceptable, so the computational mesh model with a total of 41,181,751 mesh cells was adopted. The specific mesh model is shown in Figure 3.

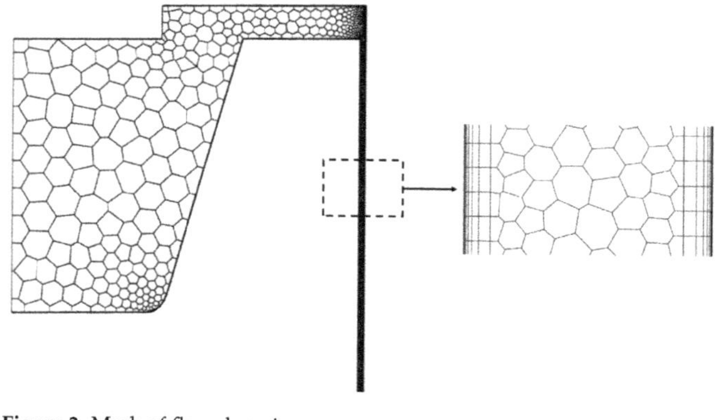

Figure 3. Mesh of flow domain.

3.2. Solver Setup

Pressure-inlet and pressure-outlet boundary conditions were applied; the inner wall (seal-ring surface) was modeled as a rotating wall with an angular speed of 7.96 rad/s, while the outer wall was stationary. A no-slip condition was enforced on all walls. A uniform axial acceleration of 9.81 m/s^2 drove the upward flow. The two radial-cut faces of the sector are defined as periodic boundaries, rotated by 10° about the z-axis.

A transient simulation was performed using the SST k-ω turbulence model. The SST k-ω model combines the advantages of the k-ε model and the k-ω model, providing higher accuracy in the near-wall region, making it especially suitable for handling turbulence characteristics within the boundary layer. Although the SST k-ω model may not capture every detail of the vortices when simulating small gap flows, LES and DNS offer higher accuracy but come with extremely high computational costs. In comparison, the SST k-ω model represents a good compromise between computational efficiency and accuracy. This is particularly important for rotating flows and complex flow scenarios. k-ω turbulence models are widely applied to study the unsteady flow field of hydraulic machinery, such as the research on the evolution of vortex rope and large curvature flow [30]. In the numerical simulation, pressure–velocity coupling was performed using the coupled method, with the PRESTO! interpolation scheme for pressure. The momentum equation and turbulence kinetic energy equation were discretized using the second-order upwind scheme to improve solution stability while ensuring accuracy. The time step was set to Δt = 0.0005 s, with 10,000 steps (total simulated time 5 s) and up to 20 inner iterations per time step.

3.3. Result

A "companion plane" is defined as the axial plane passing through the geometric center of the fluid domain within the seal-ring gap. On this plane, multiple monitoring points are arranged to record various physical quantities. Axial monitoring points point 1~point 7 (h = 0, 0.04, 0.08, 0.13, 0.16, 0.2, 0.24 m) and radial monitoring points point 4~point 4–9 (δ = 0, 0.5, 1, 1.5, 2, 2.5, 3, 3.5, 4 mm) were set. In subsequent results sections, this plane will be used to visualize key gap-flow characteristics—such as velocity and pressure—as well as rotation-induced turbulence and vortical structures. The selection of the companion plane and its schematic are shown in Figure 4.

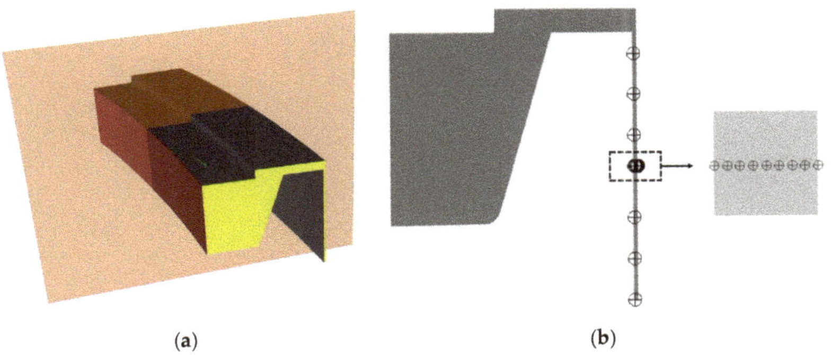

(**a**) (**b**)

Figure 4. Selection of companion plane (**a**) and schematic diagram (**b**).

3.3.1. Shear-Driven Flow Characteristics

The rotating inner wall induces a dominant circumferential velocity component, forming a shear layer within the gap. To illustrate the radial distribution of the tangential velocity magnitude, nine monitoring points were placed along a radial line on the companion plane at the inlet section height $z = 0.13$ m (see Figure 4).

Figure 5 presents the measured velocities at these nine points: the absolute velocity exhibits an inverted "S"-shaped profile. Within the inlet region of the companion plane, the tangential velocity decreases non-uniformly in the radial direction, approximately nonlinearly from 37.06 m/s at the inner wall to 0 m/s at the outer wall. The decrease is slightly steeper near the inner wall and more gradual near the outer wall, indicating mild nonlinearity. This deviation from the ideal Couette-flow profile reflects the influence of turbulent fluctuations and geometric constraints, while the overall trend confirms the leading roles of rotational driving and wall confinement on the flow field.

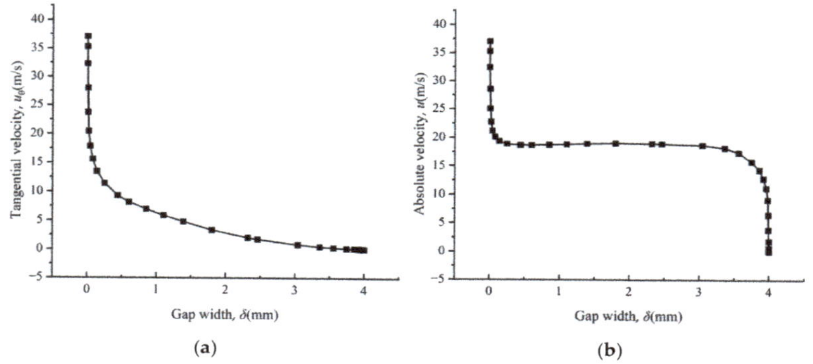

Figure 5. Speed distribution. Tangential velocity (**a**) and total velocity (**b**).

The streamline plot in Figure 6 reveals a helical, upward flow trajectory: streamlines transition from a parallel arrangement at the inlet to tightly wound coils downstream, demonstrating the superimposed effects of axial pressure driving and rotation. The helical nature of the flow is particularly pronounced near the outlet. These observations are in good agreement with the expected flow behavior.

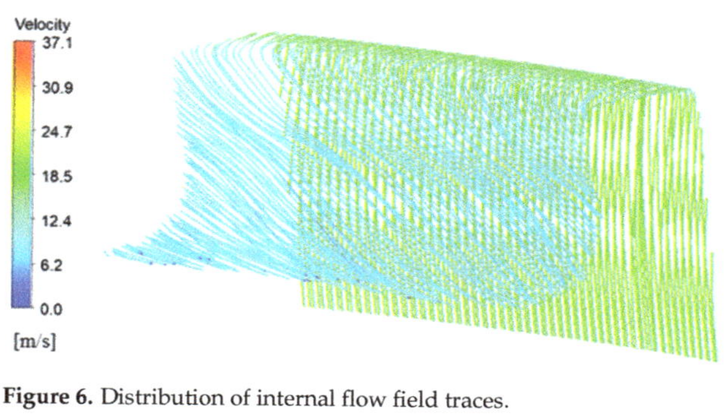

Figure 6. Distribution of internal flow field traces.

3.3.2. Influence of Short Channel Height on Flow Development

Figure 7 shows the axial pressure distribution in the inlet region: pressure decreases nearly linearly along the axis, with negligible radial variation. A localized pressure spike occurs at the top of the inlet due to geometric confinement; a distinct low-pressure region appears inside the cover cavity and at the outlet, confirming vortex formation.

Figure 8 presents axial-velocity profiles on the companion plane at three different heights. None of the velocity profiles exhibit a perfect parabolic shape, indicating the influence of the channel geometry on flow development. In the central region, the axial velocity magnitude decreases downstream, while velocities near the side walls fluctuate. Additionally, the profiles show a slight skew toward the outer wall, reflecting asymmetric development caused by the short channel height.

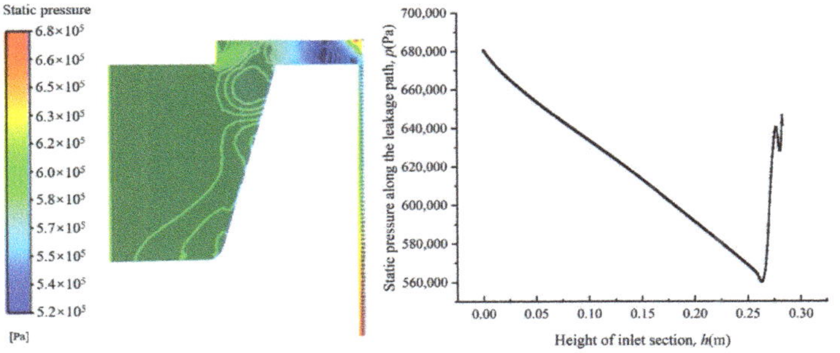

Figure 7. Pressure distribution. Static pressure distribution (**a**) and axial pressure distribution (**b**).

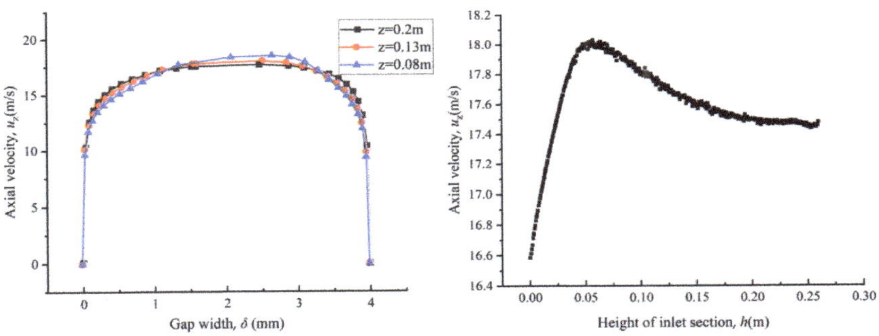

Figure 8. Axial velocity profiles at different heights of the inlet section (a) and along the centerline of the inlet section (**b**).

3.3.3. Rotation-Induced Vortices and Flow Instability

The rotating inner wall profoundly alters the gap flow, inducing complex vortical structures. Streamlines in Figure 9 illustrate spiral flow paths under rotational drive. In the inlet region, streamlines are dense and straight, indicating initial flow stability under geometric constraint. As the flow enters the cover cavity, rotational effects progressively generate vortices and helical trajectories, especially near the rotating wall. These vortices are driven by a combination of shear from the rotating wall and centrifugal forces; shear establishes velocity gradients, while centrifugal action amplifies vortex complexity.

To assess rotation-triggered instability in the inlet region, turbulent kinetic energy k was monitored at the nine radial points. Figure 10 plots k versus time for each point: the peak k of approximately 7.7 m^2/s^2 occurs adjacent to the rotating wall, indicating intense turbulent fluctuations and small-scale vortices. Turbulent kinetic energy decreases toward the channel center—forming a "U"-shaped radial profile—because the center experiences the smallest velocity gradient (hence weakest turbulence production) and greater energy dissipation. Near the outer wall, higher velocity gradients combined with centrifugally induced radial flow and turbulent diffusion elevate turbulence intensity above the central region.

3.3.4. Pressure Pulsations and Flow Instability

Transient simulations also examine the relationship between pressure pulsations and flow instability. In this paper, three characteristic points were taken on the rotating wall at the height position along the flow channel to analyze the pressure pulsation, located in the middle and top positions of the inlet, respectively. Then, fast Fourier transform was used to perform frequency-domain analysis on the pressure time–history curve (see Figure 11).

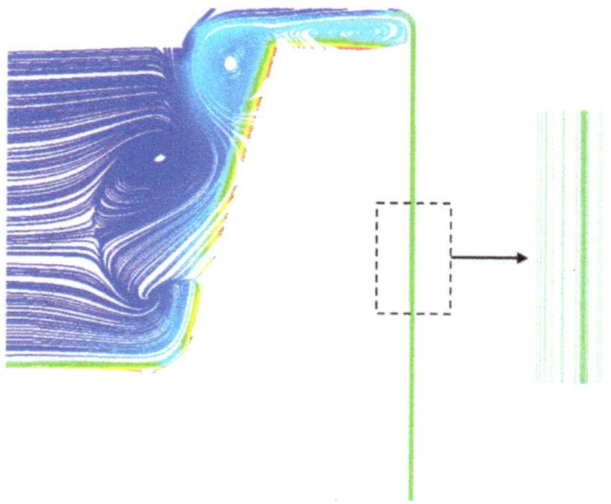

Figure 9. Flow traces in the accompanying plane.

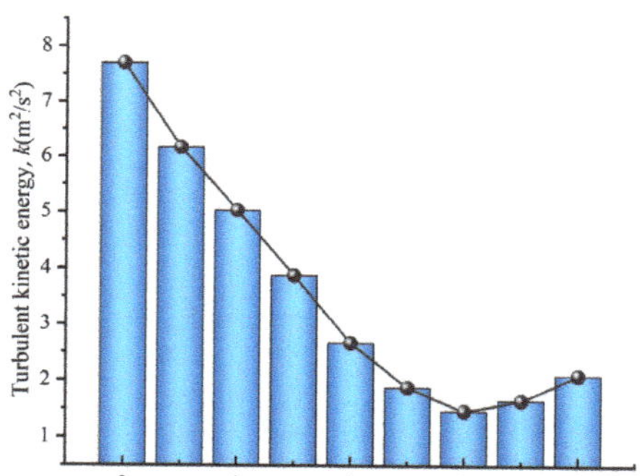

Figure 10. Changes in turbulent kinetic energy at different measuring points (point 4~4–9).

Time-series plots show that pressure fluctuation amplitudes increase along the axial direction—weak at the gap inlet but markedly stronger at the inlet–outlet interface—coinciding with the rise in turbulent kinetic energy and indicating that turbulence energy accumulation intensifies downstream instability. Frequency spectra reveal dominant pulsation frequencies in the 1.1–1.5 Hz band for all three monitoring points, closely matching the rotation frequency of 1.27 Hz. This confirms that pressure pulsations are primarily driven by the rotating inner wall. The amplitude of the dominant frequency slightly decreases from inlet to outlet, reflecting diminished low-frequency energy content due to turbulent redistribution.

3.4. Discussion

The work conducted in this study provides a visualization of the different physical parameters of the leakage ring gap. In [18], a large eddy simulation (LES) of leakage flow in step-type labyrinth seals was presented, showing the velocity distribution at the step seal position, which is consistent with the results obtained in this study. Additionally, in [31], a numerical simulation of Taylor—Couette—Poiseuille flow at Re = 10,000 was conducted using the LES model. The study focused on the effects of high rotation on the mean flow, turbulence statistics, and vortex structure. It was found that an increase in rotational speed increased the axial velocity gradient at the wall, causing the axial velocity distribution u_z in the central gap region to tend toward horizontal. The tangential velocity distribution obtained in the article aligns closely with the results of this study. Additionally, the frequency-domain analysis of the pressure fluctuations in the gap flow reveals that the frequency response of the fluctuations aligns with the existing literature [21,22] on the pulsing frequency response of labyrinth seal systems. The main frequency of the pulsations is roughly consistent with the rotational speed frequency, which reflects the accuracy of the research content.

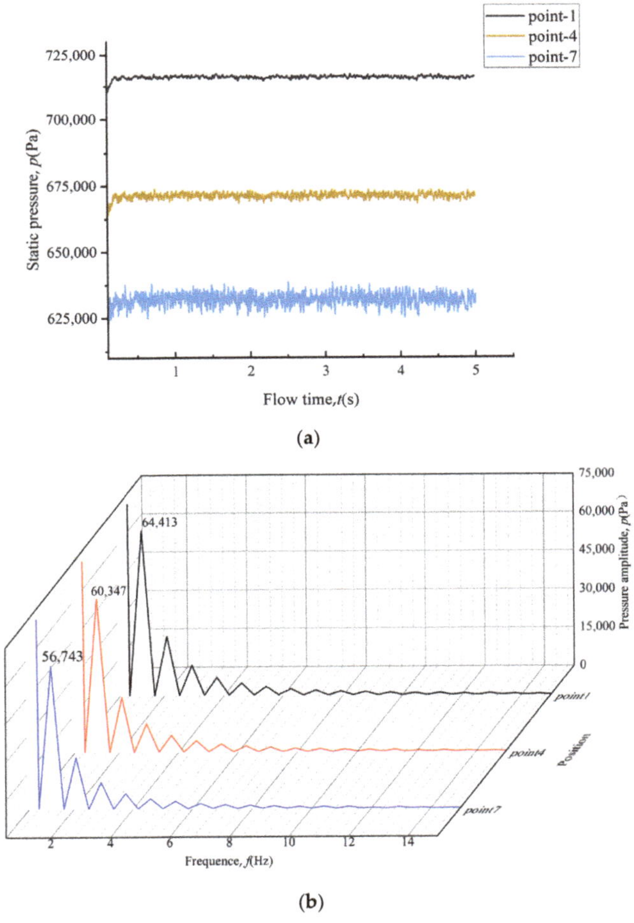

Figure 11. Pressure pulsation analysis. Pulsation time domain diagram (**a**) and pulsation frequency domain diagram (**b**).

By quantitatively comparing our results with the existing literature and providing clear explanations of boundary conditions, turbulence models, and geometric assumptions, we have not only validated the core results but have also clearly defined the scope and directions for model improvements, offering clear guidance for future research.

4. Conclusions and Outlook

4.1. Conclusions

This study systematically revealed the complex flow characteristics and multi-scale coupling mechanisms within the gap between the sealing ring and the upper frame of ultra-large mixed-flow turbines through theoretical modeling and three-dimensional transient CFD simulations. The main conclusions are as follows:

1. The CFD simulations not only validate the accuracy of the theoretical analysis but also enrich understanding of flow field characteristics through visualization. The simulation results demonstrate non-parabolic axial velocity distributions, spiral streamline trajectories, and a low-pressure zone in the upper-crown cavity, confirming the presence of vortices and the restrictive effect of short channels on flow development. The "U"-shaped distribution of turbulent kinetic energy and the dynamic variations in pressure fluctuations further reveal the rotation-induced turbulence enhancement effect, providing a powerful tool for the quantitative analysis of complex flow fields.
2. The high-speed rotation of the inner wall is the dominant factor in the flow field within the gap. The theoretical analysis derived a nearly linear distribution of tangential velocity with a shear rate as high as 9250 s^{-1}. CFD simulations further revealed an inverted "S"-shaped nonlinear distribution of tangential velocity along the radial direction, deviating from the linear characteristics of classical Couette flow. This phenomenon is attributed to the combined effects of turbulent diffusion and centrifugal force-induced secondary flows, highlighting the significant reshaping effect of rotation on boundary layer flow. Additionally, the streamline trajectories exhibited a spiral ascending flow pattern, particularly near the rotating wall, where the peak turbulent kinetic energy reaches 7.7 m^2/s^2, confirming the presence of complex vortex structures and turbulence enhancement induced by rotation. These findings provide new insights into understanding energy dissipation and leakage control in the internal flow of rotating machinery.
3. This study incorporated the multi-scale effects introduced by the short channel height and high-speed rotation into the analytical framework and employed a modified Taylor number to assess flow stability. The results indicate that although the short channel geometry suppresses the formation of complete Taylor vortices, local small-scale vortices and flow instabilities still persist, particularly near the inner wall. Spectral analysis showed that the dominant frequency of pressure fluctuations is highly correlated with the rotational frequency of 1.27 Hz, and the amplitude decreases with increasing channel height, revealing the regulatory mechanism of the coupling between rotation and geometric constraints on flow stability.

This study has demonstrated the variation trends of the flow structure and velocity distribution in the small-gap region, providing qualitative guidance for the geometric design of subsequent sealing rings. Specific structural optimization and parameter control should consider various factors such as actual operating conditions and system stability, and further research should be conducted in the future.

4.2. Future Perspectives

This study has focused on analyzing the flow dynamics characteristics within the sealing ring and upper frame gap of a super-large mixed-flow water turbine through theoretical modeling and three-dimensional transient CFD simulations. The importance and necessity of the flow characteristics in the sealing ring gap were analyzed, revealing their impact on turbine performance, including leakage, energy loss, and operational efficiency, as well as their role in ensuring safety and stability. The influence of rotating walls was examined, demonstrating how they alter velocity and pressure distributions, generating complex flow patterns such as vortices. Furthermore, turbulence and flow instability were discussed, identifying their contributions to energy dissipation and mechanical challenges such as vibration and noise. These studies highlight the complex interactions between rotation, geometry, and flow behavior in turbine sealing systems.

In the future, by leveraging data on turbulent kinetic energy, velocity gradients, and pressure pulsations obtained in this study, quantitative analyses of turbulence and instability can be conducted. Through spectral analysis, the relationship between dominant unstable frequencies and parameters such as rotational speed and gap width can be revealed. These achievements not only deepen the understanding of flow instability mechanisms but also guide the exploration of flow control strategies, such as adjusting gap geometry or introducing guide devices, to mitigate instability and reduce vibration risks, thereby enhancing the operational efficiency and safety of water turbines.

In the subsequent dynamic analysis of the sealing ring, the flow field data from this study provides precise pressure load information, laying the foundation for fluid–structure interaction analysis. Future research can apply the dynamic pressure from the CFD simulations to the finite element model of the sealing ring to calculate its stress distribution and deformation characteristics to assess whether its structural strength meets the long-term operational requirements. Additionally, by analyzing the vibration characteristics induced by pressure pulsations and combining them with material fatigue properties, it is possible to predict the fatigue life of the sealing ring, particularly focusing on potential failure risks in high-turbulence areas near the inner wall. These analytical results will provide a basis for optimizing the design of the sealing ring, such as adjusting material thickness, adding stiffeners, or optimizing geometric shapes to enhance its fatigue resistance. In summary, the flow field data from this study not only provide high-precision boundary conditions and empirical evidence for stability theoretical models and dynamic analyses but also provide guidance for optimizing the design of sealing systems, which is instrumental in significantly improving the performance and reliability of water turbines.

Author Contributions: Conceptualization, L.C.; Methodology, W.W. and Y.W.; Validation, B.X (Bing Xue). and L.X.; Formal analysis, B.X. (Baosheng Xie); Investigation, J.D.; Writing—original draft, W.W. and B.X. (Baosheng Xie); Writing—review & editing, L.C. and Y.W.; Visualization, W.W.; Supervision, Y.W.; Project administration, L.C. and Y.W. All authors have read and agreed to the published version of the manuscript.

Funding: This research was supported by the Open Research Fund of Hubei Technology Innovation Center for Smart Hydropower, Grant Number: HBCXZX-JJ-202417.

Data Availability Statement: Due to privacy and confidentiality agreements related to engineering design, detailed data supporting the results re-ported in this article, including schematics and related technical documents, cannot be made publicly available. Further inquiries can be directed to the corresponding author.

Conflicts of Interest: Authors Leilei Chen, Jian Deng, Bing Xue, and Liuming Xu were employed by China Yangtze Power Co., Ltd., Yichang. The remaining authors (Wenhao Wu, Baosheng Xie, Yuchuan Wang) declare that the research was conducted in the absence of any commercial or financial relationships that could be construed as a potential conflict of interest.

References

1. Mao, X.; Zhong, P.; Li, Z.; Liu, Z.; Hu, J.; Yan, J. Development and application of modified partially averaged Navier Stokes in the internal flow field of a Francis turbine. *Phys. Fluids* **2025**, *37*, 037122. [CrossRef]
2. International Energy Agency. *World Energy Outlook 2022*; International Energy Agency: Paris, France, 2022.
3. Rijal, M.; Neopane, H.; Chitrakar, S.; Pandey, A.; Dev, J. Numerical study of erosion on labyrinth seals of Francis turbine. *J. Phys. Conf. Ser.* **2023**, *2629*, 012021. [CrossRef]
4. Zhou, D.; Chen, H.; Zhang, J.; Jiang, S.; Gui, J.; Yang, C.; Yu, A. Numerical Study on Flow Characteristics in a Francis Turbine during Load Rejection. *Energies* **2019**, *12*, 716. [CrossRef]
5. Dörfler, P.; Sick, M.; Coutu, A. *Flow-Induced Pulsation and Vibration in Hydroelectric Machinery: Engineer's Guidebook for Planning, Design and Troubleshooting*; Springer: London, UK, 2013. [CrossRef]
6. Ohiemi, I.E.; Yang, S.; Singh, P.; Li, Y. Experimental investigation on the effect of axial gap on performance and unsteady pressure pulsations of low head axial flow hydraulic turbine. *SSRN Electron. J.* **2022**. [CrossRef]
7. Tang, B.; Liao, J.; Zhao, Z.; Deng, Q.; Li, J.; Feng, Z. Windage and Leakage Losses in Impeller Back Gap and Labyrinth Seal Cavities of Supercritical CO_2 Centrifugal Compressors. *Appl. Sci.* **2025**, *15*, 3678. [CrossRef]
8. Cui, J.; Tang, H. A review on flow instability in hydro-viscous drive. *Phys. Fluids* **2024**, *36*, 041301. [CrossRef]
9. Xie, Y.; Du, Q.; Xie, L.; Wang, Z.; Li, S. Onset of turbulence in rotor-stator cavity flows. *J. Fluid Mech.* **2024**, *991*, A3. [CrossRef]
10. Martinand, D.; Serre, E.; Lueptow, R.M. Absolute and convective instability of cylindrical Couette flow with axial and radial flows. *Phys. Fluids* **2009**, *21*, 104102. [CrossRef]
11. Nagata, M. Taylor-Couette flow in the narrow-gap Limit. *Philos. Trans. R. Soc. A* **2023**, *381*, 20220134. [CrossRef]
12. Cotrell, D.L.; McFadden, G.B. Axial flow effects on the stability of circular Couette flow with viscous heating. *Phys. Fluids* **2006**, *18*, 084106. [CrossRef]
13. Trivedi, C. Time-dependent inception of vortex rings in a Francis turbine during load variation: Large eddy simulation and experimental validation. *J. Hydraul. Res.* **2020**, *58*, 790–806. [CrossRef]
14. Trivedi, C.; Cervantes, M.J.; Dahlhaug, O.G. Numerical Techniques Applied to Hydraulic Turbines: A Perspective Review. *Appl. Mech. Rev.* **2016**, *68*, 010802. [CrossRef]
15. Moore, J.J. Three-dimensional CFD rotordynamic analysis of gas labyrinth seals. *J. Vib. Acoust.* **2003**, *125*, 427–433. [CrossRef]
16. Ding, J. Numerical Simulation and Experiment of Gas Film Flow Field on the Surface of Floating Ring Micro Groove Hydrodynamic Seal. Ph.D. Thesis, Lanzhou University of Technology, Lanzhou, China, 2019.
17. Schiffer, J.; Benigni, H.; Jaberg, H.; Schneidhofer, T.; Ehrengruber, M. Numerical simulation of the flow in a Francis turbine including the runner seals on crown and band side. *IOP Conf. Ser. Earth Environ. Sci.* **2015**, *26*, 012022. [CrossRef]
18. Kim, J.-H.; Ahn, J. Large Eddy Simulation of Leakage Flow in a Stepped Labyrinth Seal. *Processes* **2021**, *9*, 2179. [CrossRef]
19. Nicoli, A.; Johnson, K.; Jefferson-Loveday, R. Computational modelling of turbulent taylor-couette flow for bearing chamber applications: A comparison of unsteady Reynolds-averaged Navier-Stokes models. *Proc. Inst. Mech. Eng. Part A J. Power Energy* **2022**, *236*, 985–1005. [CrossRef]
20. Safari, A.; Saidi, M.H.; Salavatidezfouli, S.; Ma, W.; Yao, S. A comprehensive experimental and numerical study of Taylor-Couette flow on superhydrophobic surfaces. *AIP Adv.* **2025**, *15*, 055012. [CrossRef]
21. Jia, X.; Zhang, H.; Zheng, Q.; Fan, S.; Tian, Z. Investigation on Rotor-Labyrinth Seal System with Variable Rotating Speed. *Int. J. Turbo Jet Eng.* **2020**, *37*, 209–219. [CrossRef]
22. Zhang, X.; Jiao, Y.; Qu, X.; Huo, G.; Zhao, Z. Simulation and Flow Analysis of the Hole Diaphragm Labyrinth Seal at Several Whirl Frequencies. *Energies* **2022**, *15*, 379. [CrossRef]
23. Zheng, K.; Ma, H.; Sun, H.; Qin, J. Investigation of Flow Characteristics in Rotating Distributary and Confluence Cavities. *Energies* **2025**, *18*, 1287. [CrossRef]
24. Xu, Z.; Xu, L.; Sun, J.; Liu, M.; Liao, T.; Hu, X. Research on Micro Gap Flow Field Characteristics of Cylindrical Gas Film Seals Based on Experimental and Numerical Simulation. *Aerospace* **2024**, *11*, 40. [CrossRef]
25. Perini, M.; Binder, N.; Bousquet, Y.; Schwartz, E. Rotating Instabilities in Shrouded Low Pressure Turbine at Design and Off-Design Conditions. *J. Turbomach.* **2023**, *145*, 111006. [CrossRef]
26. Demay, Y.; Iooss, G.; Laure, P. The Couette Taylor Problem in the Small Gap Approximation. In *Ordered and Turbulent Patterns in Taylor-Couette Flow*; Andereck, C.D., Hayot, F., Eds.; NATO ASI Series; Springer: Boston, MA, USA, 1992; Volume 297. [CrossRef]

27. Munson, B.R.; Young, D.F.; Okiishi, T.H.; Huebsch, W.W. *Fundamentals of Fluid Mechanics*, 8th ed.; Wiley: Hoboken, NJ, USA, 2018; ISBN 978-1-119-40787-3.
28. Taylor, G.I. The Stability of a Viscous Liquid Contained Between Two Rotating Cylinders. *Philos. Trans. R. Soc. Lond. A* **1923**, *223*, 289–343. [CrossRef]
29. Chandrasekhar, S. *Hydrodynamic and Hydromagnetic Stability*; Dover Publications: New York, NY, USA, 1981; p. 24.
30. Mao, X.; Liu, Z.; Li, T.; Mao, G.; Chen, D. A brief review of numerical solving methods for internal fluid of pumped storage unit. *Int. J. Energy Res.* **2020**, *44*, 7549–7566. [CrossRef]
31. Gavrilov, A.; Ignatenko, Y. Numerical Simulation of Taylor-Couette-Poiseuille Flow at Re = 10,000. *Fluids* **2023**, *8*, 280. [CrossRef]

energies

MDPI

Article

Research on the Influence of Impeller Oblique Cutting Angles on the Performance of Double-Suction Pumps

Zhongsheng Wang [1], Xinxin Li [2], Jun Liu [1], Ji Pei [2,*], Wenjie Wang [2], Kuilin Wang [2] and Hongyu Wang [2]

[1] Shandong Shuanglun Co., Ltd., Weihai 264203, China
[2] National Research Center of Pumps, Jiangsu University, Zhenjiang 212013, China; wenjiewang@ujs.edu.cn (W.W.)
* Correspondence: jpei@ujs.edu.cn

Abstract

Double-suction centrifugal pumps are extensively employed in industrial applications owing to their high efficiency, low vibration, superior cavitation resistance, and operational durability. This study analyzes how impeller oblique cutting angles (0°, 6°, 9°, 12°) affect a double-suction pump at a fixed 4% trimming ratio and constant average post-trim diameter. Numerical simulations and tests reveal that under low-flow ($0.7Q_d$) and design-flow conditions, the flat-cut (0°) minimizes reflux ratio and maximizes efficiency by aligning blade outlet flow with the mainstream. Increasing oblique cutting angles disrupts this alignment, elevating reflux and reducing efficiency. Conversely, at high flow ($1.3Q_d$), the 12° bevel optimizes outlet flow, achieving peak efficiency. Pressure pulsation at the volute tongue (P11) peaks at the blade-passing frequency, with amplitudes significantly higher for 9°/12° bevels than for 0°/6°. The flat-cut suppresses wake vortices and static–rotor interaction, but oblique cutting angle choice critically influences shaft-frequency pulsation. Entropy analysis identifies the volute as the primary loss source. Larger oblique cutting angles intensify wall effects, increasing total entropy; pump chamber losses rise most sharply due to worsened outlet velocity non-uniformity and turbulent dissipation. The flat-cut yields minimal entropy at Q_d. These findings provide a basis for tailoring impeller trimming to specific operational requirements. Furthermore, the systematic analysis provides critical guidance for impeller trimming strategies in other double-suction pumps and pumps as turbines in micro hydropower plants.

Keywords: double-suction pump; impeller oblique cutting; identification of vortex; pressure pulsation; energy losses

1. Introduction

Fluid machinery, as one of the primary energy-consuming equipment types in industrial sectors, requires performance optimization to achieve energy conservation. Among these systems, double-suction centrifugal pumps are extensively utilized in diverse industrial applications—including water conservancy irrigation systems, municipal water supply and drainage, industrial water circulation, and inter-basin water transfer projects—due to their high efficiency, low vibration, excellent cavitation resistance, ease of maintenance, and prolonged service life [1]. However, operational scenarios frequently involve mismatches, where pump performance parameters exceed actual demand. To address this, impeller trimming has become a prevalent engineering solution for adjusting hydraulic characteristics and expanding operational adaptability.

Impeller trimming remains a prevalent technique for centrifugal pump performance adjustment, with extensive research conducted on trimming laws and their hydraulic impacts. In [2], the authors investigated multiple trimming configurations, revealing distinct efficiency variations across operating conditions. A survey in [3] demonstrated that front shroud modifications significantly affect low-specific-speed pump performance. In [4], the authors established that increased tip clearance through trimming reduces head while amplifying vibrations. In [5], the authors identified nonlinear deviations from affinity laws regarding radial forces and pressure pulsations under trimmed diameters. A survey optimized operational efficiency through controlled trimming, though excessive cuts degraded performance [6]. In [7], the authors observed shifted efficiency zones and reduced pressure pulsations in self-priming pumps post-trimming. An investigation reported a 1.7% prolonged self-priming duration with a diameter reduction of 6% [8]. In [9], the authors documented the best efficiency point migration toward lower flows with trimming. A survey in [10] identified turbine-mode efficiency gains through partial trimming. In [11], the authors achieved pressure fluctuation mitigation via trailing-edge modifications. In [12], the authors reduced axial forces through strategic back-shroud trimming. A survey minimized hydraulic impacts on multistage impellers through optimized trimming [13]. A survey demonstrated that oblique trimming alleviates rotating stall in inter-blade channels [14].

In engineering practice, the initial cutting amount is typically determined by referencing the standard trimming law. However, derived under idealized operating assumptions, this law's calculated results often exhibit certain deviations. To mitigate impeller damage risks from excessive trimming, a multi-stage trimming approach is universally adopted. This involves achieving target dimensions through two or more incrementally staged trimming operations, with mandatory hydraulic performance testing after each trimming procedure to evaluate compliance with predetermined performance indicators.

The primary limitation of this traditional progressive trimming method lies in its repetitive "trimming–testing" cycles, which not only consume substantial labor hours but also significantly increase economic costs—particularly pronounced during trimming operations for pumps with higher specific speeds.

To address these deficiencies and provide engineering guidance for impeller trimming technology, a survey established a corrective calculation diagram for optimizing centrifugal pump and fan impeller diameters through systematic analysis of experimental data [15]. However, subsequent research revealed limitations in the applicability of correction coefficients due to significant variations in critical geometric parameters (including hub-to-diameter ratio and blade number) within test samples. Furthermore, the computational accuracy of this correction method deteriorates with increasing the specific speed of turbomachinery, exhibiting more significant discrepancies particularly in high-specific-speed pumps.

Current research on internal flow characteristics of double-suction pumps post-cutting remains limited, with existing studies predominantly focusing on external performance parameters and pressure pulsations [16], while insufficient attention has been given to transient flow phenomena such as vortex structures, recirculation patterns, and jet–wake interactions. Compared to their single-suction counterparts, double-suction pumps exhibit greater structural complexity, where impeller trimming significantly alters internal flow dynamics. The selection of optimal cutting angles critically influences energy loss mitigation, operational longevity, and maintenance cost reduction. Consequently, in-depth investigation into the influence of impeller oblique cutting on the hydraulic performance of double-suction pumps holds significant implications for subsequent engineering design and energy retrofit projects. Furthermore, within the context of low-carbon transition initiatives, research on impeller cutting for double-suction pumps proves critically important when these pumps operate in turbine mode.

As illustrated in Figure 1, this study employs high-resolution numerical simulations to systematically investigate the influence of impeller oblique cutting on double-suction pump performance. The research comprehensively examines three interconnected aspects: (1) the evolution patterns of external characteristics (head, efficiency) with varying oblique cutting angles, (2) the underlying mechanisms governing internal flow modifications including vortex dynamics and energy dissipation, and (3) the resulting pressure pulsation phenomena particularly at critical monitoring locations. This tripartite analysis bridges macro-scale performance degradation with micro-scale flow instabilities, providing holistic insights for hydraulic optimization.

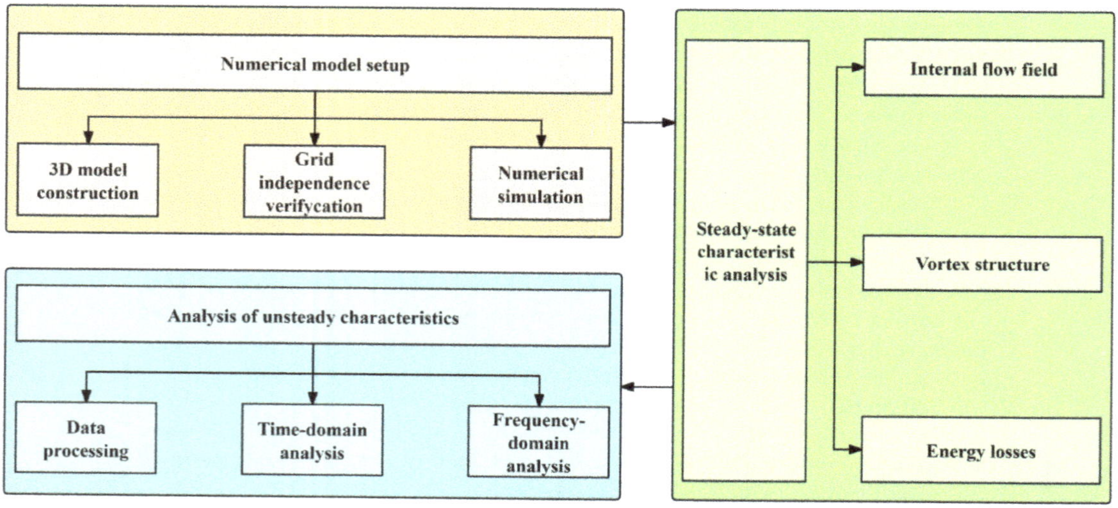

Figure 1. The flowchart of the research.

2. Model and Numerical Simulation

2.1. Physical Model

The investigation focuses on a 250GS40 horizontal split-case single-stage double-suction pump manufactured by Shandong Shuanglun Co., Ltd., (Weihai, China). The computational domain comprises four principal components: semi-spiral suction chamber, pump chambers, impeller passages, and volute flow channels. Considering the recirculation effects at discharge, a 5D extension pipe (D = outlet diameter) was appended to the volute outlet to minimize boundary condition interference, with detailed structural configuration presented in Figure 2. The prototype operates at design parameters Q_d = 500 m^3/h, H = 40 m, and rotational speed n = 1480 rpm, yielding specific speed n_s = 90. Complete geometric specifications are cataloged in Table 1.

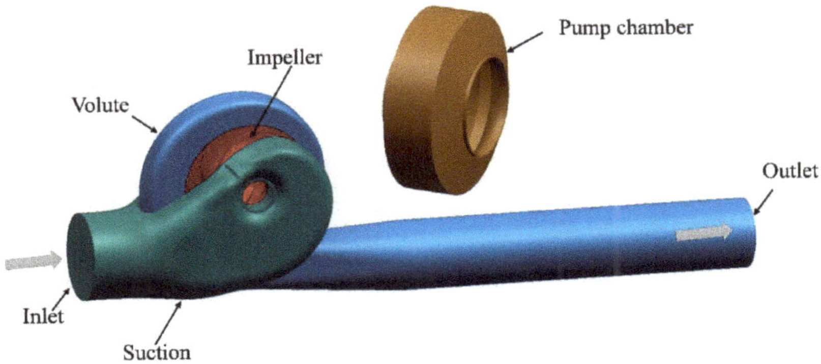

Figure 2. Schematic diagram of the structure of a double-suction pump.

Table 1. Main structural parameters.

Parameters	Value
Specific speed n_s	90
Impeller suction diameter $D_1/(\text{mm})$	192
Impeller outlet diameter $D_2/(\text{mm})$	365
Impeller inlet install angle $\beta_1/(°)$	20.8
Impeller outlet install angle $\beta_2/(°)$	29.6
Impeller outlet width $b_2/(\text{mm})$	46
Number of blades Z	6
Rotational speed $n/(\text{r/min})$	1480
Volute inlet width $b_3/(\text{mm})$	100

2.2. Impeller Cutting Scheme

Forward oblique cutting of double-suction impeller outlets involves symmetrical material removal from the front shroud to the back shroud under constant post-trimming mean diameter D_2', creating distinct angular configurations (θ), as shown in Figure 3. The geometric parameters are defined as follows: trimming angle θ, original diameter D_2, trimmed mean diameter D_2', shroud/hub diameters D_{21} and D_{22}, trimming allowance ΔD, and trimming ratio φ_D. The governing relationships are as follows:

$$\Delta D = D_2 - D_2' \tag{1}$$

$$\varphi_D = \frac{\Delta D}{D_2} \tag{2}$$

$$D_2' = \frac{(D_{21} + D_{22})}{2} \tag{3}$$

$$0 < \theta < \arctan\frac{2\Delta D}{b_2} \tag{4}$$

Table 2 presents recommended allowable cutting ranges for impellers across specific speeds. Extensive engineering practice and numerical analyses demonstrate an inverse correlation between permissible trimming ratios and specific speeds, whereby higher n_s values correspond to reduced allowable trimming percentages. Practical implementations require parameter adjustments based on operational demands and performance targets. Within permissible trimming limits, this study adopts a 4% trimming ratio while maintaining constant post-cutting mean diameter. Four angular configurations—0° (planar cutting),

6° (oblique cutting), 9° (oblique cutting), and 12 (oblique cutting)—were systematically designed for comparative analysis.

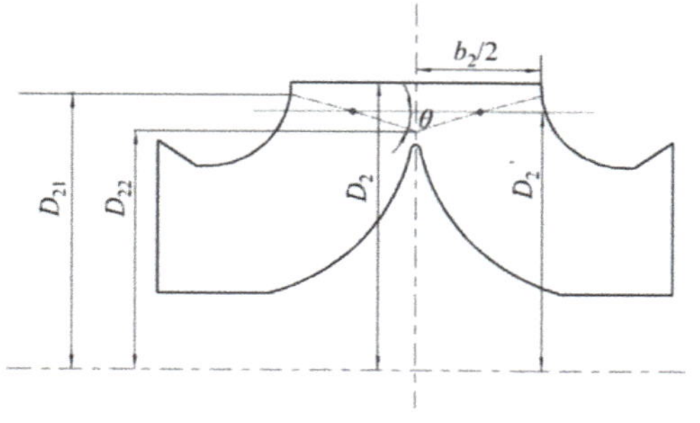

Figure 3. Schematic diagram of impeller cutting.

Table 2. Specific speed and allowable cutting allowance.

n_s	60	120	200	300	500
φ_D	0.2	0.15	0.11	0.09	0.07

2.3. Grid Independence Verification

To systematically determine the optimal grid resolution balancing computational accuracy and efficiency, a progressive refinement approach was implemented for the double-suction pump's 3D flow passage model, as illustrated in Figure 4. To accurately capture near-wall flow characteristics, the mesh near wall surfaces of flow components was refined with boundary layer grids. Based on geometric requirements, five inflation layers were generated at an expansion ratio of 1.2. Grid dependency analysis evaluated pump efficiency, head, and shaft power across multiple mesh densities, ultimately selecting 6.43 million elements through convergence verification (Table 3). This methodology ensures computational accuracy while effectively managing resource expenditure. Three prototype configurations underwent grid independence validation under identical block topology structures to eliminate geometric bias. The consistent meshing strategy maintained uniform node distribution parameters across all trimming configurations, mitigating flow comparison distortions from grid artifacts.

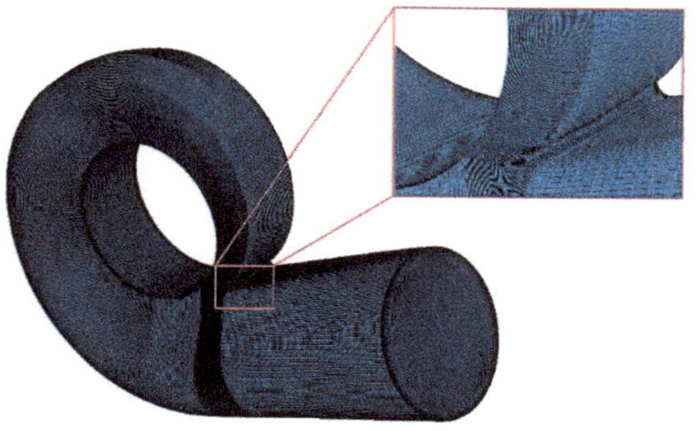

(**a**)

Figure 4. *Cont.*

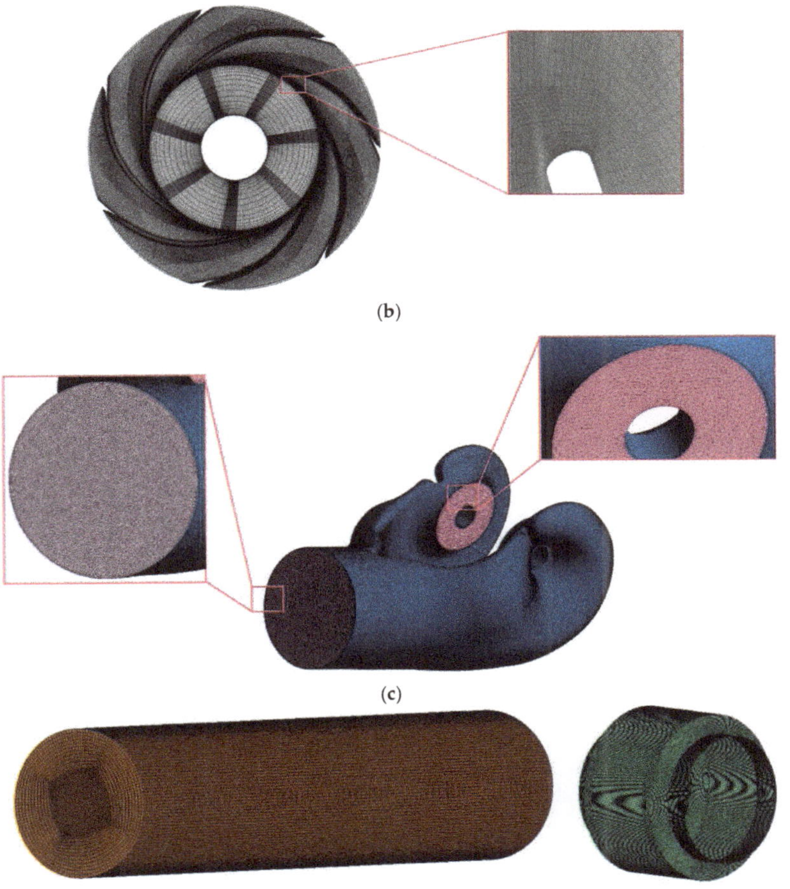

(b)

(c)

(d) (e)

Figure 4. Schematic diagram of the grid of each component. (**a**) Mesh of volute. (**b**) Mesh of impeller. (**c**) Mesh of suction chamber. (**d**) Mesh of outlet pipe. (**e**) Mesh of pump chamber.

Table 3. Verification results of grid independence.

Specific Speed	Number of Meshes/$\times 10^6$	Efficiency/%
$n_s = 90$	2.28	83.31
	5.54	85.66
	6.43	85.87
	7.86	85.87
	8.21	85.88

2.4. Numerical Simulation

Numerical simulations were performed using the ANSYS CFX 2020 R2, with the SST k-ω turbulence model selected for this study. Compared to the k-ε and Standard k-ω models, the SST k-ω model demonstrates superior capability in capturing complex flow features within centrifugal pumps, exhibiting closer agreement with experimental flow characteristics [17]. The SST k-ω model enhances the standard k-ω formulation by modifying the definition of eddy viscosity to account for the transport of principal turbulent shear stresses. This improvement enables more accurate predictions of adverse pressure gradient flows. The turbulent kinetic energy k and specific dissipation rate ω are governed by Equation (5) and Equation (6), respectively.

$$\frac{\partial(\rho k)}{\partial t} + \frac{\partial(\rho k u_i)}{\partial x_i} = P_k - \beta^* \rho k \omega + \frac{\partial}{\partial x_j}\left[(\mu + \sigma_k \mu_t)\frac{\partial k}{\partial x_j}\right] \tag{5}$$

$$\begin{aligned} \frac{\partial(\rho\omega)}{\partial t} + \frac{\partial(\rho\omega\mu_i)}{\partial z_i} &= \alpha\frac{\omega}{k}P_k - \beta\rho\omega^2 + \frac{\partial}{\partial z_j}\left[(\mu + \sigma_\omega \mu_t)\frac{\partial \omega}{\partial x_j}\right] \\ &+2(1-F_1)\rho\sigma_{\omega 2}\frac{1}{\omega}\frac{\partial k}{\partial x_j}\frac{\partial \omega}{\partial x_j} \end{aligned} \tag{6}$$

where P_k represents the turbulent kinetic energy production term; μ_i denotes the turbulent viscosity; α, β, β^*, σ_k, σ_ω, $\sigma_{\omega 2}$ are model-specific constants; and F1 is the blending function.

Numerical simulations were conducted using ANSYS CFX to analyze the full-flow field characteristics of the prototype pump. The fluid medium was defined as water at 25 °C, with turbulence modeling employing the SST k-ω formulation. Boundary conditions comprised a total pressure inlet (reference pressure: 1 atm) and mass flow outlet. The computational domain was partitioned into rotating (impeller) and stationary (pump chamber, volute, suction chamber) regions, interconnected via interface boundaries. For steady-state simulations, the Frozen Rotor method governed data transfer at rotor–stator interfaces, with solution convergence achieved when RMS residuals reached below 10^{-6}.

Unsteady simulations utilized steady-state results as initial conditions, implementing Transient Frozen Rotor interface treatment with General Grid Interface (GGI) connectivity. The temporal resolution was configured with a 3° impeller rotation per timestep (0.000337838 s), maintaining a maximum of 10 iterations per step. The simulation duration encompassed 14 full impeller revolutions (0.567568 s), preserving the convergence criterion of RMS residuals 10^{-6}.

2.5. Experimental Verification

Experimental verification was conducted on an open test rig at Shandong Shuanglun Co., Ltd. to validate numerical simulation accuracy. The prototype pump was instrumented with identical pressure sensors at both inlet and outlet, while electromagnetic flow sensors measured volumetric flow rates. Critical sensor specifications are detailed in Table 4, with the test bench configuration illustrated in Figure 5.

Table 4. Main parameters of the sensors.

Parameter	Sensor Type	Measurement Accuracy	Range
Flow rate	Electromagnetic flow transducer	±0.3%	0–1000 m^3/h
Pressure	EJA530E Pressure sensor	±0.1%	0–1 MPa

Figure 6 presents comparative performance curves between numerical simulations and experimental measurements for the prototype pump. Steady-state simulations employed time-averaged results from the final 120 iterations, demonstrating close agreement with test data: maximum relative errors of 1.9% in efficiency and 2.1% in head—both within acceptable 3% thresholds—confirm the computational model's reliability in capturing internal flow dynamics. Observed discrepancies originate primarily from turbulence model limitations and experimental measurement uncertainties.

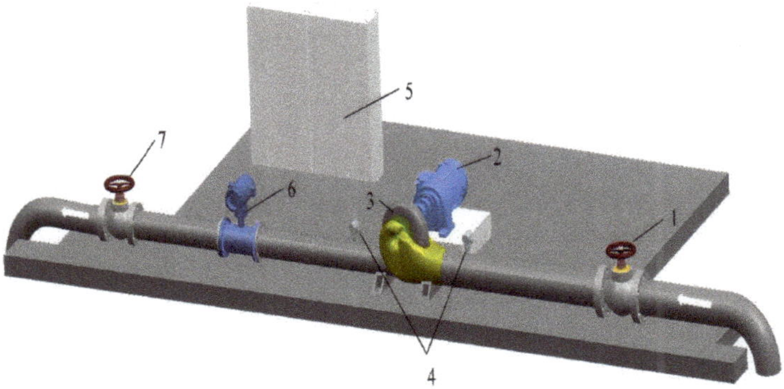

Figure 5. Layout of the test rig.

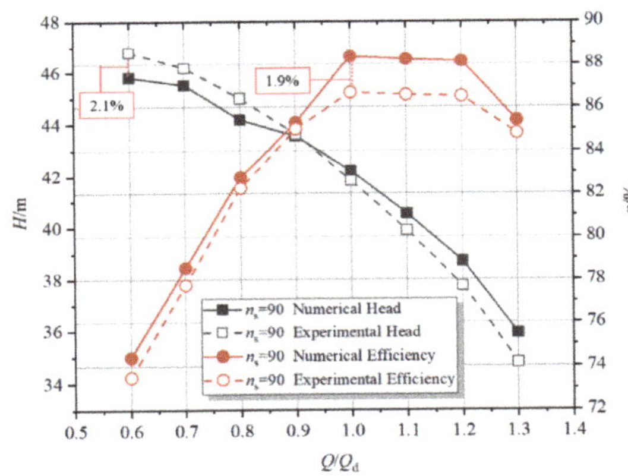

Figure 6. Comparison of external characteristic curves between the simulated and experimental.

3. Results and Discussion

3.1. The Influence of the Impeller Oblique Cutting Angle on the Flow at the Impeller Outlet

In centrifugal pumps, fluid typically flows radially outward from the impeller inlet to outlet. When vortex components emerge in the meridional plane velocity distribution, partial fluid reverses direction against the primary flow, a phenomenon termed flow reversal. The intensity of this reversal reflects flow disorder degree and energy dissipation magnitude. The primary flow rate Q and reversed flow rate Q_{rev} through cross-section S can be calculated using the following equations [18]

$$Q = \int_S v_{\text{n}} \text{d}S \quad (7)$$

$$Q_{\text{rev}} = \int_S \frac{|v_{\text{n}}| - v_{\text{n}}}{2} \text{d}S \quad (8)$$

where v_{n} represents the normal-direction velocity component at cross-section S, with flow direction convention defined as positive for outward radial movement (impeller inlet to outlet) and negative for inward flow reversal.

Focusing on the impeller outlet where oblique cutting occurs, this investigation evaluates flow pattern alterations through reversed flow percentage *R*, calculated using:

$$R = \frac{Q_{\text{rev}}}{Q} \times 100\% \quad (9)$$

where Q_{rev} denotes reversed flow rate, and *Q* represents total through flow.

The influence of impeller oblique cutting angles on outlet reflux ratios and pump efficiency across operational conditions is illustrated in Figure 7. Analysis reveals a consistent decrease in reflux ratios with increasing flow rates for all configurations, peaking at partial-load conditions ($0.7Q_d$) due to reduced axial velocities inducing negative incidence angles at blade inlets. This flow separation mechanism, driven by adverse pressure gradients, generates localized reflux zones at the impeller outlet. At $1.3Q_d$, diminished pressure gradients suppress reverse flow tendencies, resulting in reduced reflux ratios.

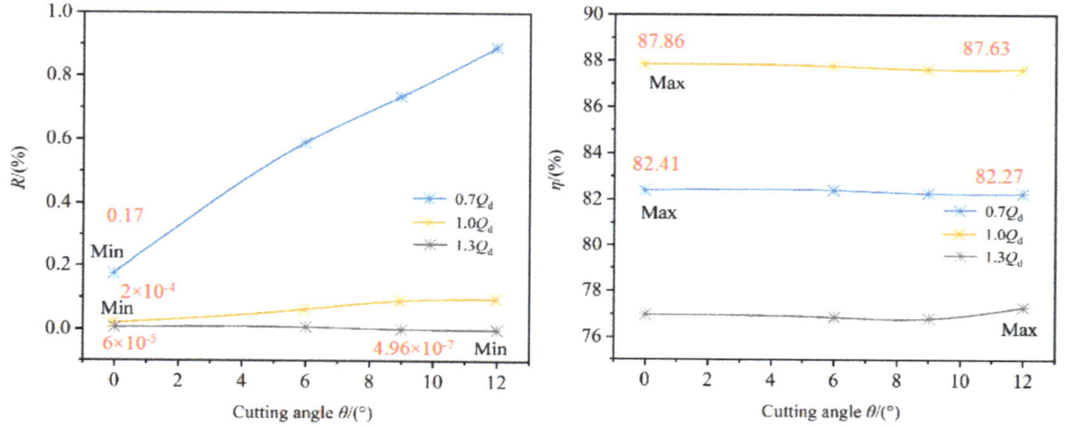

Figure 7. Comparison of reflux ratio at the impeller outlet and efficiency with different oblique cutting angles under various working conditions.

Under partial-load conditions ($0.7Q_d$), reflux ratios exhibit significant positive correlation with cutting angles (θ), while efficiency demonstrates marginal inverse dependence ($\Delta\eta < 1\%$). Similar trends emerge at design condition, with reflux ratios increasing and efficiency decreasing slightly as θ enlarges. Conversely, high-flow operation ($1.3Q_d$) shows minimal reflux ratio variation ($\theta = 12°$ yielding optimal reduction) and non-monotonic efficiency behavior—minimum at $\theta = 9°$ then recovering to peak at $\theta = 12°$.

Comparative evaluation confirms the baseline planar cut ($\theta = 0°$) achieves minimal reflux ratios and maximum efficiency at both partial-load and design conditions. This superiority originates from optimal alignment between blade exit angles and primary flow direction, minimizing flow deviation. Oblique cutting induces progressive flow misalignment, elevating reflux ratios and reducing hydraulic efficiency. Notably, moderate oblique angles ($\theta = 12°$) enhance high-flow performance through improved outlet flow guidance and turbulence mitigation.

To investigate the influence of different oblique cutting angles on the flow field within the impeller, the X-Y cross-section of the impeller, specifically the meridional plane, was constructed as shown in Figure 8. The coordinate origin is located at the geometric center of the impeller, with the inflow direction aligned with the X-axis.

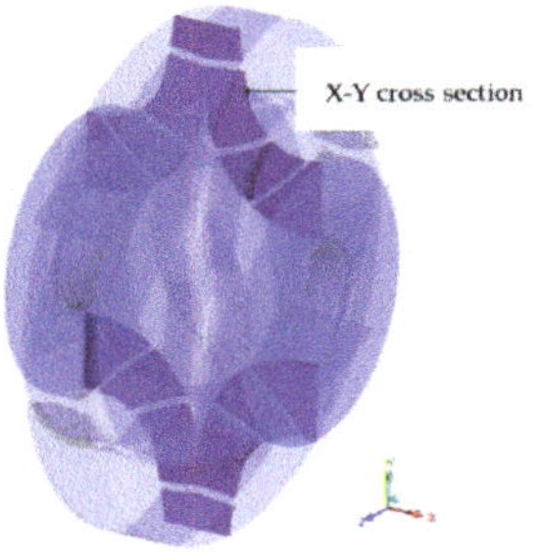

Figure 8. The X-Y cross-section position of the impeller.

The analysis of the velocity streamlines on the X-Y cross-section for impellers with different oblique cutting angles (Figure 9) clearly reveals significant flow instability characteristics within the impeller. Across the operating range from $0.7Q_d$ to $1.3Q_d$, the flow velocity in the main flow zone exhibits an increasing trend with rising flow rate for all bevel configurations, reaching its maximum value at the impeller outlet.

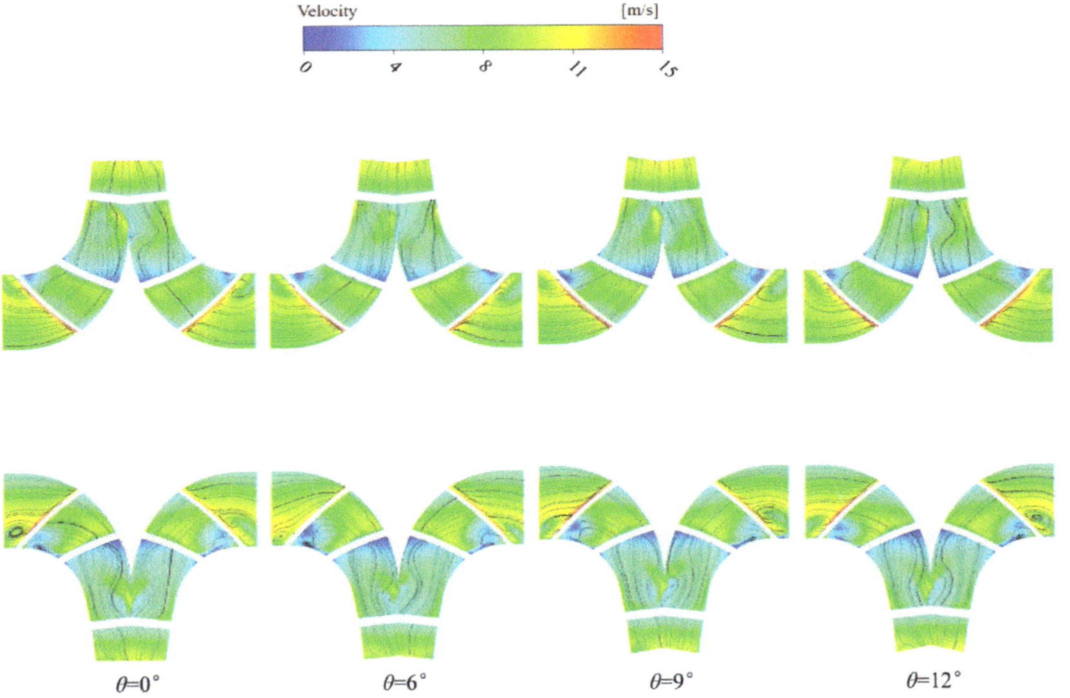

(**a**)

Figure 9. *Cont.*

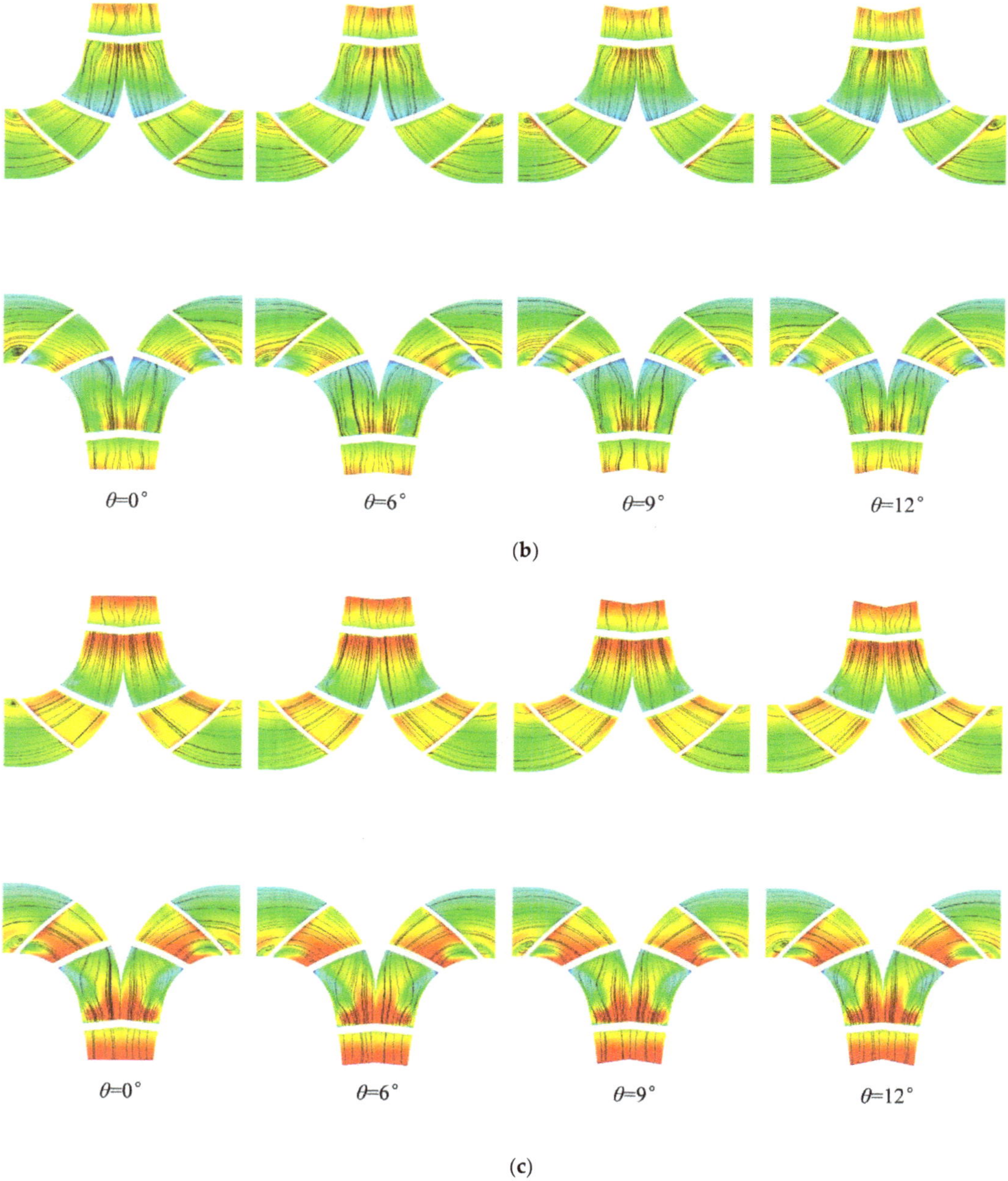

Figure 9. The velocity streamline distribution on the X-Y cross-section of the impeller with different oblique cutting angles under various working conditions. (**a**) $0.7Q_d$. (**b**) $1.0Q_d$. (**c**) $1.3Q_d$.

Under design condition and high-flow conditions, the flow field distribution for each oblique cutting angles scheme is relatively uniform. Small reflux zones appear near the impeller inlet in all cases. At the design flow rate, the concentration of streamlines slightly intensifies as the oblique cutting angles increases.

Conversely, under low-flow conditions, all four oblique cutting angles schemes exhibit inlet reflux and vortex zones. Some streamlines are densely packed and exhibit significant curvature. Notably, the inlet vortex is more pronounced in the flat-cut impeller. This indicates that beveling the impeller can improve the flow conditions in the inlet region under low-flow operating conditions.

3.2. The Influence of the Impeller Oblique Cutting Angles on the Vortex Structure

A systematic investigation into the impeller oblique cutting angle's impact on vortex structures in double-suction pumps was conducted using the Ω vortex identification method. This approach enables precise discrimination of rotational motions and concurrent detection of vortices across intensity scales. By implementing a 0.52 threshold [19–21], vortices with distinct strength levels were differentiated within the impeller. Spatial-temporal evolution patterns of vortex structures were visualized through turbulent kinetic energy-based color mapping (as shown in Figure 10).

Under design conditions, three characteristic vortex types were consistently observed across four oblique cutting angle configurations: Leading Edge Vortices (LEV) from blade-leading-edge flow separation, Wake Vortices (WV) induced by trailing-edge jet–wake interactions, and Tip Leakage Vortices (TLV) originating from blade–tip boundary layer separation. Progressive increases in oblique cutting angles modified the blade exit angle, redirecting fluid flow paths. This redirection amplified suction-side low-pressure zones near trailing edges, driving sequential expansion of WV, LEV, and TLV regions. The spatial growth of these vortices correlated with elevated energy dissipation via intensified turbulent interactions. Three-dimensional streamline visualizations depict vortex structure distributions within the impeller, as presented in Figure 11. Progressive extension of low-velocity zones toward the blade leading edge regions is observed with increasing oblique cutting angles, while simultaneously intensified flow disorder emerges near the suction surface trailing edge. This progressive flow variation directly visualizes alterations in impeller–internal vortex dynamics.

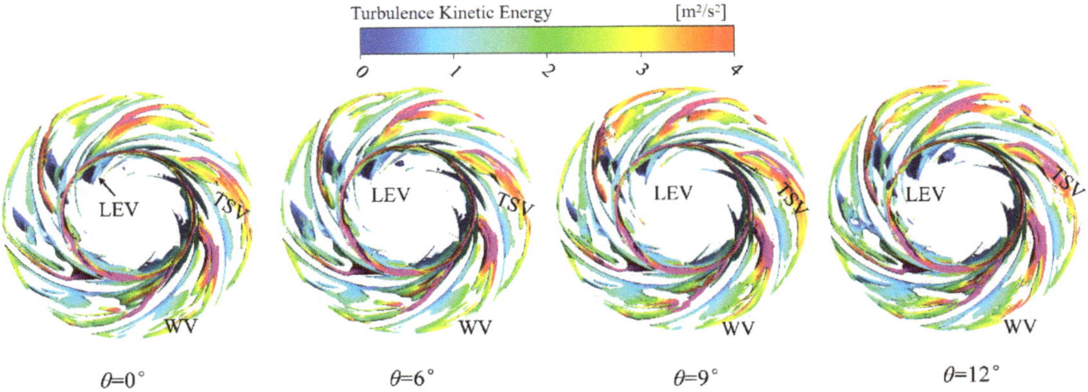

(a)

Figure 10. *Cont.*

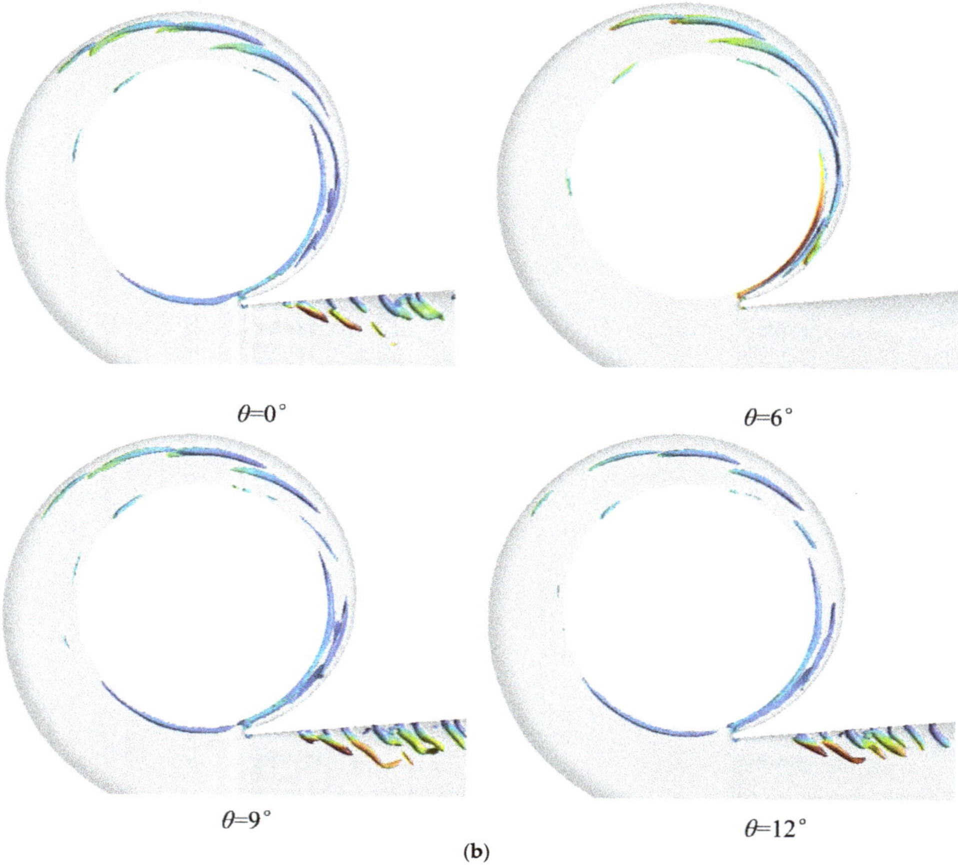

Figure 10. Comparison of vortex structures distributions of impeller and volute with different oblique cutting angles. (**a**) Impeller. (**b**) Volute.

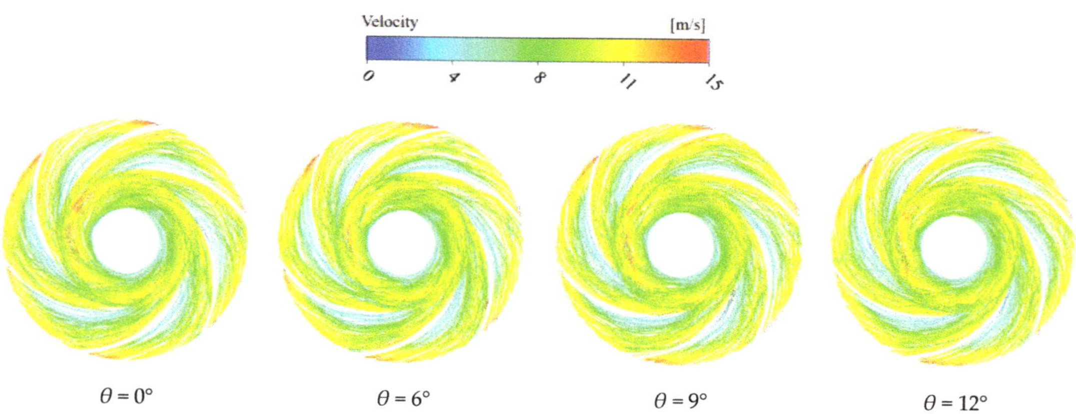

Figure 11. Comparison of 3D velocity streamline distribution of impeller with different oblique cutting angles.

Volute analysis identified transient vortices in the diffuser section, signaling flow instabilities. While vortex density increased with larger oblique cutting angles, a critical transition occurred at 6°: diffuser vortices vanished abruptly, coinciding with amplified turbulent kinetic energy at the volute inlet. This phenomenon suggests a threshold-dependent shift in energy transfer mechanisms, where classical vortex dissipation pathways are supplanted by alternative loss modes.

3.3. The Influence of the Impeller Oblique Cutting Angle on Pressure Pulsation

To investigate the transient pressure field characteristics induced by impeller oblique cutting in double-suction pumps, systematically positioned pressure pulsation monitoring points were implemented within the flow domain, as illustrated in Figure 12. Eight monitoring points (P1–P8) were uniformly distributed along the volute casing at 45° intervals, while three monitoring points (P9–P11) were concentrated near the volute tongue to capture flow separation characteristics. Additional monitoring points were positioned within the impeller flow channels (M1–M2), pump cavities (M3–M4), blade pressure surfaces (M7–M8), and blade suction surfaces (M5–M6).

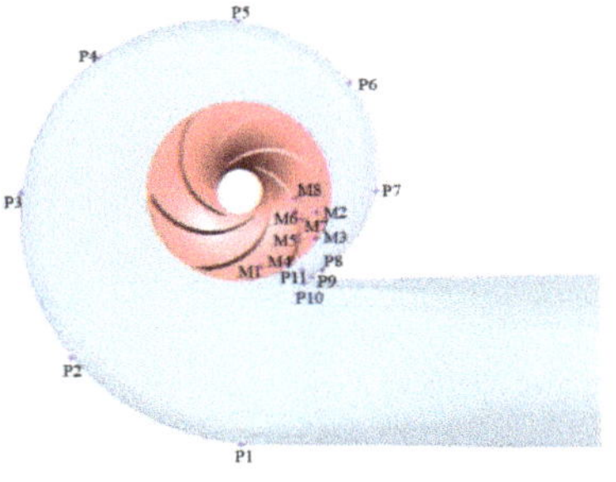

Figure 12. Distribution of pressure monitoring points.

As shown in Figure 13, pressure variations at monitoring point P11 exhibited periodic behavior across three rotational cycles for all oblique cutting angle configurations, with 18 peaks and 18 troughs corresponding to the impeller's six blades. This confirms the blade-induced periodic modulation of pressure fluctuations near the volute tongue. Pressure fluctuation amplitudes for the 9° and 12° bevel configurations exceeded those of the flat-cut and 6° oblique cutting. Collectively, these results demonstrate that flat-cut or smaller oblique cutting angles reduce static-rotor interaction intensity and subsequently suppress pulsation amplitudes at the volute tongue.

Frequency–domain analysis of pressure data (Figure 14) under design conditions revealed dominant spectral peaks at f_{BPF} for monitoring points P3, P6, P8, and P11, indicating blade-passing effects as the primary excitation source. At monitoring points P3, the 9° oblique cutting configuration exhibited the lowest amplitude at f_{BPF} but higher amplitude at f_{n} compared to other configurations, suggesting a trade-off between frequency components during blade-edge modification. Monitoring points P8 and P11 near the volute tongue displayed pronounced harmonic frequencies and periodic fluctuations, reflecting strong static–rotor interaction. Notably, the flat-cut impeller generated weaker f_{BPF} amplitudes at P11 compared to oblique cutting configurations, indicating suppressed wake vortices formation and reduced flow instability.

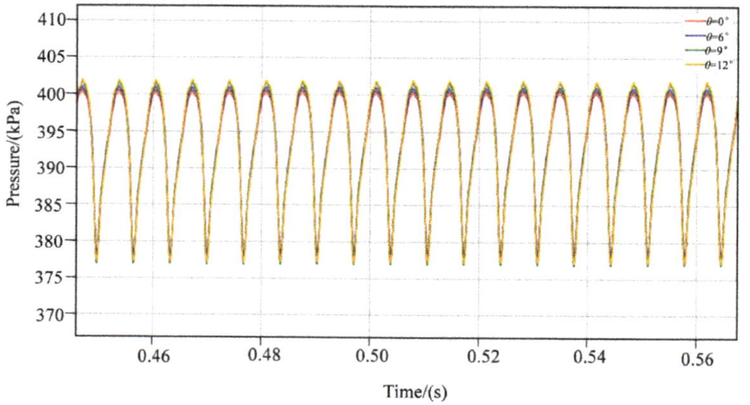

Figure 13. Time-domain pressure pulsation at monitoring point P11 under different operating conditions with varying oblique cutting angles.

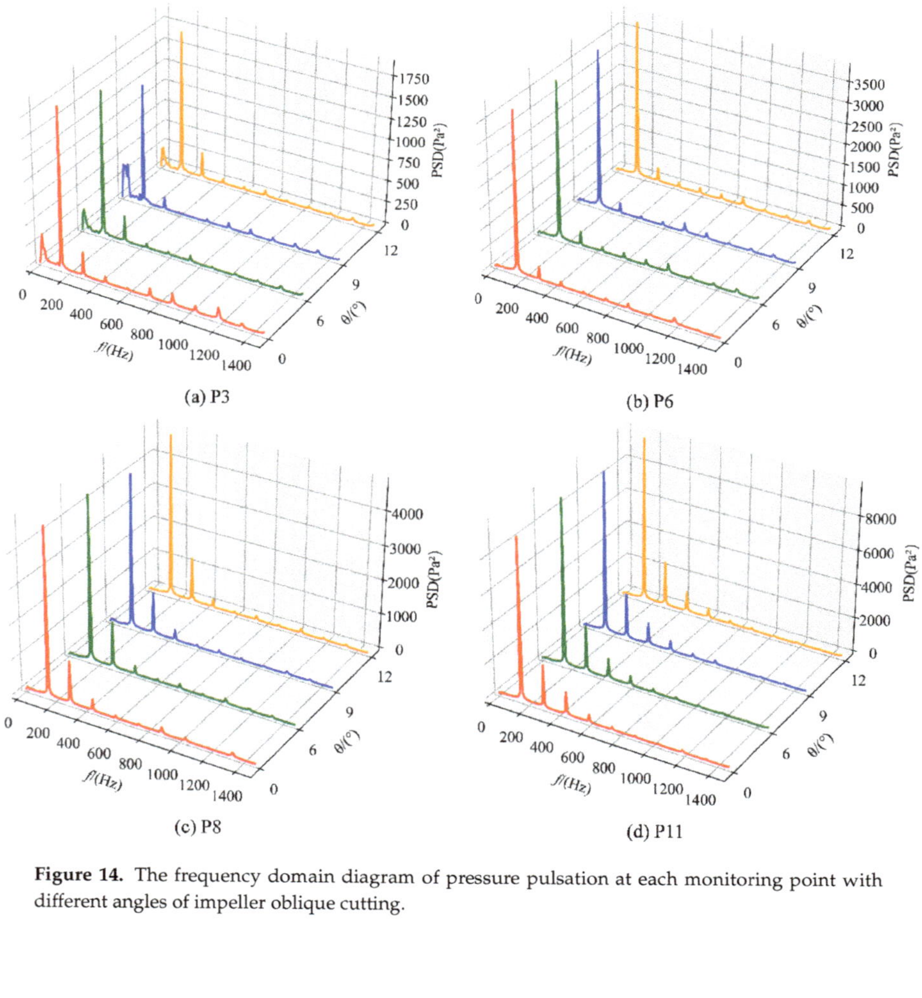

Figure 14. The frequency domain diagram of pressure pulsation at each monitoring point with different angles of impeller oblique cutting.

These results demonstrate that oblique cutting angle selection significantly affects both blade-passing frequency (f_{BPF}) and shaft frequency (f_{n}) pressure fluctuations. The flat-cut impeller exhibited superior performance in mitigating wake vortices and static–rotor interaction, providing critical insights for optimizing impeller trimming strategies in double-suction pumps.

3.4. The Influence of the Impeller Oblique Cutting Angles on Energy Loss

According to the second law of thermodynamics, for any irreversible process, its entropy will always increase. Under the operating conditions of the double suction pump, the mechanical energy of the impeller is converted into the kinetic energy of the fluid. During this energy conversion process, due to the existence of viscous stress, Reynolds stress, and turbulence in the fluid during operation, an irreversible conversion process from mechanical energy to internal energy is inevitably triggered, which is manifested as an increase in the entropy value of the system, that is, the entropy generation phenomenon. Therefore, the entropy generation theory can be applied to numerical simulation to evaluate the energy loss of double-suction pumps.

In the turbulent flow of a single-phase incompressible ideal fluid, the entropy transport equation can be expressed as

$$\rho\left(\frac{\partial s}{\partial t}+u\frac{\partial s}{\partial x}+v\frac{\partial s}{\partial y}+w\frac{\partial s}{\partial z}\right)=-\mathrm{div}\left(\frac{\vec{q}}{T}\right)+\frac{\Phi}{T}+\frac{\Phi_{\Theta}}{T^2} \tag{10}$$

where ρ represents the fluid density, s represents the specific entropy parameter, the velocity field is composed of Cartesian components u, v, and w, and T characterizes the thermodynamic temperature. The right side of the equation contains three key terms: reversible heat exchange term $-\mathrm{div}\left(\frac{\vec{q}}{T}\right)$, viscous dissipation entropy generation term $\frac{\Phi}{T}$, and heat transfer temperature difference entropy generation term $\frac{\Phi_{\Theta}}{T^2}$.

According to the Reynolds mean theory, the flow variable is decomposed into the sum of the time-mean quantity and the pulsation quantity. The entropy transport control equation obtained through the Reynolds mean operation can be expressed as

$$\rho\left(\frac{\partial \overline{s}}{\partial t}+\overline{u}\frac{\partial \overline{s}}{\partial x}+\overline{v}\frac{\partial \overline{s}}{\partial y}+\overline{w}\frac{\partial \overline{s}}{\partial z}\right)=\overline{\mathrm{div}\left(\frac{\vec{q}}{T}\right)}-\rho\left(\frac{\partial \overline{u's'}}{\partial x}+\frac{\partial \overline{v's'}}{\partial y}+\frac{\partial \overline{w's'}}{\partial z}\right)+\frac{\overline{\Phi}}{T}+\frac{\overline{\Phi_{\Theta}}}{T^2} \tag{11}$$

where $\overline{\mathrm{div}\left(\frac{\vec{q}}{T}\right)}-\rho\left(\frac{\partial \overline{u's'}}{\partial x}+\frac{\partial \overline{v's'}}{\partial y}+\frac{\partial \overline{w's'}}{\partial z}\right)$ represents the reversible heat transfer amount. The internal flow of hydraulic machinery is an approximately adiabatic process, and the entropy production rate $\frac{\overline{\Phi_{\Theta}}}{T^2}$ due to the increase in the heat transfer temperature difference can be ignored. Therefore, the entropy production rate $\frac{\overline{\Phi}}{T}$ due to the increase in dissipation can be directly obtained.

$$\frac{\overline{\Phi}}{T}=\dot{S}'''_{\overline{D}}+\dot{S}'''_{D'} \tag{12}$$

where $\dot{S}'''_{\overline{D}}$ is the entropy production rate caused by the average velocity, usually used to represent direct dissipation, and $\dot{S}'''_{D'}$ is the entropy production rate caused by the pulsating velocity, usually used to represent turbulent dissipation [22,23].

$$\dot{S}_{\overline{D}}''' = \frac{\mu_{eff}}{\overline{T}}\left\{2\left[\left(\frac{\partial\overline{u}}{\partial x}\right)^2+\left(\frac{\partial\overline{v}}{\partial y}\right)^2+\left(\frac{\partial\overline{w}}{\partial z}\right)^2\right]+\left(\frac{\partial\overline{u}}{\partial y}+\frac{\partial\overline{v}}{\partial x}\right)^2+\left(\frac{\partial\overline{u}}{\partial z}+\frac{\partial\overline{w}}{\partial x}\right)^2+\left(\frac{\partial\overline{v}}{\partial z}+\frac{\partial\overline{w}}{\partial y}\right)^2\right\} \tag{13}$$

$$\dot{S}_{D'}''' = \frac{\mu}{\overline{T}}\left\{2\left[\overline{\left(\frac{\partial u'}{\partial x}\right)^2}+\overline{\left(\frac{\partial v'}{\partial y}\right)^2}+\overline{\left(\frac{\partial w'}{\partial z}\right)^2}\right]+\overline{\left(\frac{\partial u'}{\partial y}+\frac{\partial v'}{\partial x}\right)^2}+\overline{\left(\frac{\partial u'}{\partial z}+\frac{\partial w'}{\partial x}\right)^2}+\overline{\left(\frac{\partial v'}{\partial z}+\frac{\partial w'}{\partial y}\right)^2}\right\} \tag{14}$$

$$\mu_{eff} = \mu + \mu_t \tag{15}$$

where μ is dynamic viscosity, μ_{eff} is effective dynamic viscosity, and μ_t is turbulent dynamic viscosity.

It is known that speed and temperature can be used to determine $\dot{S}_{\overline{D}}'''$, while $\dot{S}_{D'}'''$ cannot be solved directly. When the SST k-ω turbulence model is selected, $\dot{S}_{D'}'''$ can be expressed as

$$\dot{S}_{D'}''' = \alpha\frac{\rho\omega k}{T} \tag{16}$$

where $\alpha = 0.09$, and ω and k are the turbulent vortex frequency and turbulence intensity, respectively. Integrating the local entropy production rate of the calculation domain can obtain the entropy production caused by the time-average velocity and the pulsating velocity [24,25], and the formula is as follows:

$$S_{pro,\overline{D}} = \int_V \dot{S}_{\overline{D}}''' dV \tag{17}$$

$$S_{pro,D'} = \int_V \dot{S}_{D'}''' dV \tag{18}$$

In addition, the velocity gradient and pressure gradient on the surface of the centrifugal pump blade are relatively high, and there is a strong wall effect in the flow field. The entropy generated by the wall effect cannot be ignored, and its calculation formula is

$$S_{pro,W} = \int_A \frac{\vec{\tau}\cdot\vec{v}}{T} dA \tag{19}$$

where $\vec{\tau}$ represents the shear stress of the wall surface, and $\vec{v}$ represents the velocity at the first layer grid near the wall surface.

To sum up, the total entropy production within the calculation domain is

$$S_{pro} = S_{pro,\overline{D}} + S_{pro,D'} + S_{pro,W} \tag{20}$$

Figure 15 shows the influence law of the model pump at different oblique cutting angles under different working conditions on the total entropy production. It can be seen from the figure that with the increase in flow rate, the total entropy production of each oblique cutting scheme shows a trend of first decreasing and then increasing. Meanwhile, under both the low-flow condition and the design condition, the total entropy production increases slowly with the increase in the angle, but its increase rate shows a nonlinear relationship with the oblique cutting angles of the impeller. Under the working condition of high flow rate, with the increase in the cutting angle, the total entropy production in the pump shows a change characteristic of first slightly increasing and then significantly decreasing. Especially when the impeller is beveled by 12°, the total entropy production decreases significantly, indicating that under the working condition of high flow rate, adopting a larger impeller oblique cutting angles can effectively reduce the total entropy production of the double-suction pump. Therefore, in the optimal design of double-suction pumps, the reasonable selection of the impeller oblique cutting angles is one of the key parameters for the performance of the lift pump.

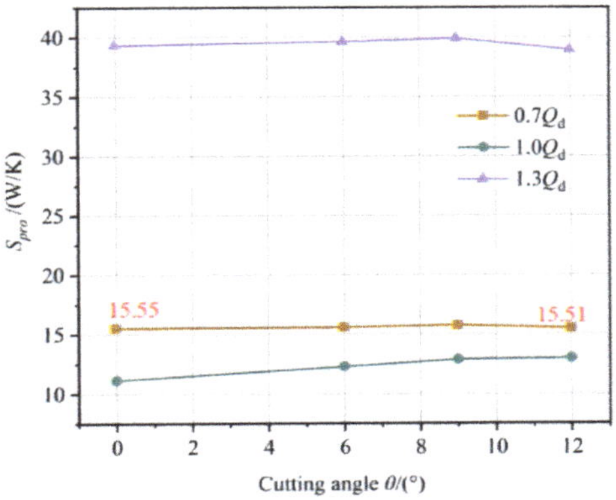

Figure 15. The influence law of the pump being beveled at different angles under different working conditions on the total entropy production.

To further analyze the relationship between the impeller oblique cutting angle and the entropy output of each type, Figure 16 compares the variations in the impeller oblique cutting angle on the entropy production of the three types under the design conditions of the model pump. Since the direct dissipation proportion of each scheme is approximately 0.31% to 0.46%, always less than 0.3%, the wall dissipation caused by the wall effect and the turbulent dissipation caused by the pulsating velocity are analyzed with emphasis. As can be seen from the figure, in all four beveling schemes, wall dissipation dominates, accounting for approximately 58.94% to 63.10% of the total entropy production. Therefore, when calculating the influence of impeller beveling on the energy loss of the double-suction pump, the impact of the avoidance effect on energy loss cannot be ignored, and wall loss must be taken into account. With the increase in the slant angle, the turbulent dissipation loss slightly increases, but the change is not obvious. Meanwhile, with the increase in the slant angle, the wall dissipation of each slant scheme gradually increases. This indicates that for this model pump, the increase in the slant angle will intensify the wall effect, resulting in more energy loss. Figure 17 shows the influence law of the total entropy generation of each flow component of the model pump when the impeller is beveled at different angles at the designed flow rate. From the analysis of the diagram, it can be seen that the entropy loss of each flow-through component is in the order of volute, impeller, pump chamber, and suction chamber. The volute has always been the main flow-through component for the total entropy generation of each model, with its proportion ranging from 41.66% to 50.17%. However, the entropy generation contribution of the suction chamber is the smallest, with its proportion always less than 5%, and it is less affected by the change in the impeller oblique cutting angle. Under the design working conditions, as the oblique cutting angle increases, the total entropy production in the pump shows a continuous upward trend. Among them, the loss in the pump cavity increases most significantly, and its increase rate is basically consistent with the growth rate of the total entropy production. When the impeller is horizontally cut, the total entropy production reaches the minimum value of 11.16 W/K. For the model pump, the beveling of the impeller has the greatest impact on the entropy loss in the pump cavity, and its proportion increases from 15.62% to 27.58%. This indicates that an increase in the beveling angle of the impeller may cause the velocity distribution of the fluid at the blade outlet to tend to be uneven, and increase the complexity of the fluid passing through the transition area between the blade and the pump cavity. Uneven velocity distribution and complex geometric shape changes are prone to causing turbulence, especially inside the pump cavity. The formation of turbulence will lead to a large amount of kinetic energy being converted into thermal energy, thereby increasing energy loss.

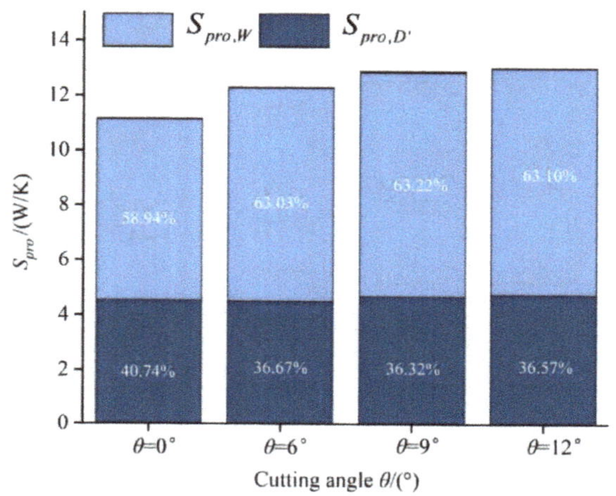

Figure 16. Comparison of entropy production of various types at different oblique cutting angles under the design working conditions.

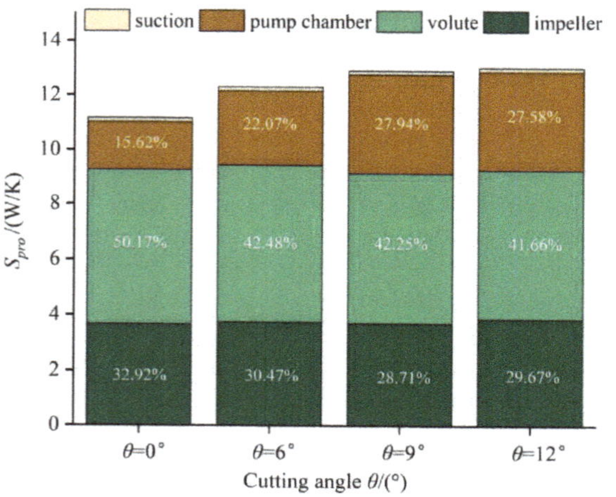

Figure 17. Comparison of entropy production of each flow component at different oblique cutting angles under the design working conditions.

From the above analysis, it is clear that different angles of the impeller's oblique cutting have the greatest impact on entropy loss in the pump cavity. In order to further analyze the energy loss in the transition zone between the impeller and the pump cavity caused by the impeller's oblique cutting angle, a visual analysis of the losses at the impeller and the pump cavity is conducted. It can be seen from Figure 18 that the entropy generation is mainly concentrated on the suction surface of the blade and the junction area between the impeller outlet and the pump cavity. In the case of a flat cut of the impeller, the energy loss caused by reflux on the suction surface of the blade is relatively large. As the angle increases to 6°, the loss at the suction surface of the blade slightly decreases. As the angle increases, the range of the high-entropy production area in the pump cavity gradually expands, indicating that the entropy production loss in the pump cavity also increases accordingly. Simultaneously, combined with the vortex structure distribution above, it can be found that the high entropy generation region also corresponds to the high turbulent kinetic energy region. This spatial correspondence further substantiates that vortex formation inherently induces quantifiable entropy production.

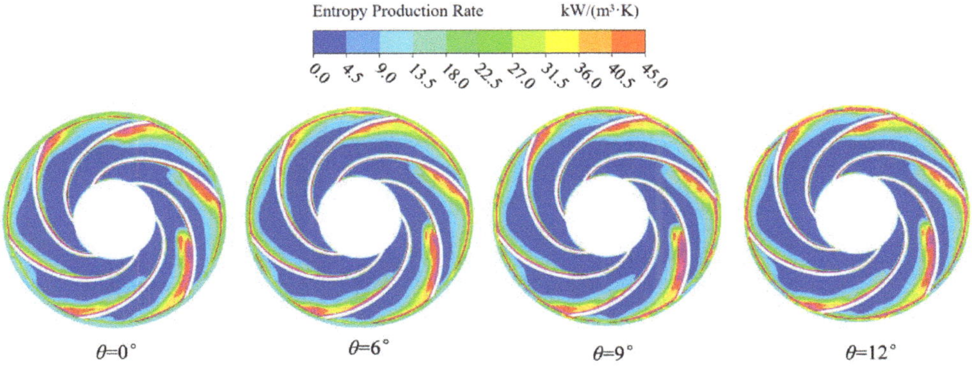

Figure 18. The entropy production rate distribution of impellers at various oblique cutting angles and the cross-section of the pump cavity.

4. Conclusions

This study focused on a double-suction centrifugal pump. Under the condition of ensuring the same average outer diameter, the influence of different oblique cutting angles (0°, 6°, 9°, 12°) on the flow characteristics inside the pump was systematically compared through a combination of numerical simulation and experiments. The specific conclusions obtained are as follows:

(1) When the impeller is cut horizontally under low-flow conditions and design conditions, the direction of the blade outlet is consistent with the main flow direction, and the fluid can flow out of the impeller at the optimal angle. As the oblique cutting angles increases, the flow at the blade outlet begins to deviate from the ideal direction, resulting in an increase in the reflux ratio and a decrease in efficiency. Under the working condition of high flow rate, the reflux ratio at the impeller outlet slightly decreases with the increase in the oblique cutting angles. It is the smallest when the impeller is bevel at 12°, and the efficiency is also the largest when it is bevel at 12°. This indicates that under this working condition, bevel at a certain angle can improve the outlet situation at the impeller outlet and enhance the fluidity of the fluid.

(2) At monitoring point P11 in the volute tongue area, the amplitude of pressure pulsation is the largest, showing a distinct feature dominated by blade frequency. Moreover, the pulsation period corresponds to the number of blades. The amplitudes of pressure pulsation in the 9° and 12° oblique cutting schemes are significantly higher than those

in the flat cutting and 6° oblique cutting schemes. The selection of the oblique cutting angles not only affects the intensity of the blade frequency, but also has a significant impact on the pressure pulsation at the shaft frequency. However, for the flat cutting of the impeller of this double-suction pump, it shows obvious advantages in suppressing the formation of the wake vortices and reducing the static–rotor interaction. These findings provide an important reference basis for the optimization of the impeller cutting of the double-suction pump.

(3) This double-suction pump mainly relies on wall dissipation. An increase in the oblique cutting angles will intensify the wall effect, leading to an increase in energy loss and an increase in total entropy production. The volute is always the component with the maximum entropy generation, accounting for 41.66% to 50.71%. The bevel cutting of the impeller has the greatest impact on the pump cavity loss. Under the design conditions, the total entropy generation of the flat cutting scheme is the lowest. The bevel cutting will increase the non-uniformity of the impeller outlet velocity and intensify the turbulence loss in the pump cavity.

(4) While the current study employs the SST k-ω turbulence model, future work could implement high-fidelity approaches such as Large Eddy Simulation (LES) or Detached Eddy Simulation (DES). These advanced methodologies would enhance the resolution of complex turbulent structures and transient flow characteristics induced by impeller beveling, particularly improving prediction accuracy for oblique cutting-modified wake vortices and static–rotor interaction effects.

(5) Given that pumps predominantly operate at the design point, this study prioritizes the design operating condition while maintaining invariant key geometric parameters such as blade count. Future investigations could examine the post-beveling performance of double-suction pumps across varying specific speeds and blade counts.

Author Contributions: Conceptualization, W.W.; Methodology, Z.W.; Validation, J.L.; Formal analysis, J.P.; Investigation, X.L.; Data curation, H.W.; Visualization, K.W. All authors have read and agreed to the published version of the manuscript.

Funding: The study was supported by the National Natural Science Foundation of China (Grant No. 52209112) and the National Key Research and Development Program of China (Grant No. 2022YFC3202901).

Data Availability Statement: All the data in this paper are obtained by physical experiment and numerical simulation, and the data used to support the findings of this study are available from the corresponding author upon request.

Conflicts of Interest: Authors Zhongsheng Wang and Jun Liu were employed by the company Shandong Shuanglun Co., Ltd. The remaining authors declare that the research was conducted in the absence of any commercial or financial relationships that could be construed as a potential conflict of interest.

References

1. Miao, S.; Tan, X.; Luo, W.; Wang, X.; Yang, J. The mechanism of internal energy losses in double-suction centrifugal pumps under direct and reverse conditions. *Energy* **2024**, *306*, 132547. [CrossRef]
2. Qu, X.; Wang, L. Effects of impeller trimming methods on performances of centrifugal pump. *J. Energy Eng.* **2016**, *142*, 04016008. [CrossRef]
3. Shadab, M.; Karimipour, M.; Najafi, A.F.; Paydar, R.; Nourbakhsh, S.A. Effect of impeller shroud trimming on the hydraulic performance of centrifugal pumps with low and medium specific speeds. *Eng. Appl. Comput. Fluid Mech.* **2022**, *16*, 514–535. [CrossRef]

4. Khalifa, A.E. Performance and vibration of a double volute centrifugal pump: Effect of impeller trimming. In *ASME International Mechanical Engineering Congress and Exposition*; American Society of Mechanical Engineers: New York, NY, USA, 2014; Volume 46476, p. V04AT04A063. [CrossRef]
5. Gonzalez, J.; Parrondo, J.; Santolaria, C.; Blanco, E. Steady and unsteady radial forces for a centrifugal pump with impeller to tongue gap variation. *J. Fluids Eng.* **2006**, *128*, 454–462. [CrossRef]
6. Šavar, M.; Kozmar, H.; Sutlović, I. Improving centrifugal pump efficiency by impeller trimming. *Desalination* **2009**, *249*, 654–659. [CrossRef]
7. Qiu, G.; Zhu, S.; Wang, K.; Wang, W.; Hu, J.; Hu, Y.; Zhi, X.; Qiu, L. Numerical study on the dynamic process of reciprocating liquid hydrogen pumps for hydrogen refueling stations. *Energy* **2023**, *281*, 128303. [CrossRef]
8. Wang, K.; Zhang, Z.; Jiang, L.; Liu, H.; Li, Y. Effects of impeller trim on performance of two-stage self-priming centrifugal pump. *Mech. Eng.* **2017**, *9*, 1687814017692493. [CrossRef]
9. Han, Y.; Li, H.; Tiganik, T.; Wang, Y.; Zhou, L. Influence mechanism of trimming impeller diameter in a centrifugal pump by computational fluid dynamics investigation. *J. Fluids Eng.* **2023**, *145*, 021205. [CrossRef]
10. Jain, S.V.; Swarnkar, A.; Motwani, K.H.; Patel, R.N. Effects of impeller diameter and rotational speed on performance of pump running in turbine mode. *Energy Convers. Manag.* **2015**, *89*, 808–824. [CrossRef]
11. Huang, B.; Zeng, G.; Qian, B.; Wu, P.; Shi, P.; Qian, D. Pressure fluctuation reduction of a centrifugal pump by blade trailing edge modification. *Processes* **2021**, *9*, 1408. [CrossRef]
12. Shi, W.D.; Zhou, L.; Lu, W.G.; Li, H. Numerical simulation and experimental study on deep-well centrifugal pump with small rear shroud diameter. *Mater. Res.* **2012**, *354*, 659–663. [CrossRef]
13. Xu, Z.; Zhang, F.; Zhu, L.; Chen, K.; Zhang, J.; Hong, Q. Effects of impeller Trimming on energy performance and fluid-structure interaction characteristics of the two-stage pump. *Energy* **2025**, *317*, 134698. [CrossRef]
14. Ye, W.; Zhuang, B.; Wei, Y.; Luo, X.; Wang, H. Investigation on the unstable flow characteristic and its alleviation methods by modifying the impeller blade tailing edge in a centrifugal pump. *J. Energy Storage* **2024**, *86*, 111358. [CrossRef]
15. Stepanoff, A.J. *Centrifugal and Axial Flow Pumps: Theory, Design, and Application*; ASCE: Reston, VA, USA, 1991.
16. Yates, M.A.; Weybourne, I. Improving the energy efficiency of pumping systems. *J. Water Supply Res. Technol. AQUA* **2001**, *50*, 101–111. [CrossRef]
17. Ramadhan Al-Obaidi, A. Effects of different turbulence models on three-dimensional unsteady cavitating flows in the centrifugal pump and performance prediction. *Int. J. Nonlinear Sci. Numer. Simul.* **2019**, *20*, 487–509. [CrossRef]
18. Wang, K.; Ju, Y.; Zhang, C. A quantitative evaluation method for impeller-volute tongue interaction and application to squirrel cage fan with bionic volute tongue. *J. Fluids Eng.* **2019**, *141*, 081104. [CrossRef]
19. Tian, S.; Gao, Y.; Dong, X.; Liu, C. Definitions of vortex vector and vortex. *J. Fluid Mech.* **2018**, *849*, 312–339. [CrossRef]
20. Zhang, F.; Appiah, D.; Chen, K.; Yuan, S.; Adu-Poku, K.A.; Wang, Y. Dynamic characterization of vortex structures and their evolution mechanisms in a side channel pump. *J. Fluids Eng.* **2020**, *142*, 111502. [CrossRef]
21. Tran, C.T.; Long, X.; Ji, B. Vortical structures in the cavitating flow in the Francis-99 draft tube cone under off-design conditions with the new omega vortex identification method. *J. Phys.* **2019**, *1296*, 012011. [CrossRef]
22. Kock, F.; Herwig, H. Local entropy production in turbulent shear flows: A high-Reynolds number model with wall functions. *Int. J. Heat Mass Transf.* **2004**, *47*, 2205–2215. [CrossRef]
23. Kock, F.; Herwig, H. Entropy production calculation for turbulent shear flows and their implementation in CFD codes. *Int. J. Heat Fluid Flow* **2005**, *26*, 672–680. [CrossRef]
24. Herwig, H.; Kock, F. Direct and indirect methods of calculating entropy generation rates in turbulent convective heat transfer problems. *Heat Mass Transf.* **2007**, *43*, 207–215. [CrossRef]
25. Herwig, H.; Kock, F. Local entropy production in turbulent shear flows: A tool for evaluating heat transfer performance. *J. Therm. Sci.* **2006**, *15*, 159–167. [CrossRef]

energies

MDPI

Article

Development of a Cavitation Indicator for Prediction of Failure in Pump-As-Turbines Using Numerical Simulation

Maciej Janiszkiewicz [1] and Aonghus McNabola [2,*]

[1] Department of Civil, Structural and Environmental Engineering, Trinity College Dublin, D02 PN40 Dublin, Ireland; janiszkm@tcd.ie
[2] School of Engineering, RMIT University, 124 La Trobe Street, Melbourne, VIC 3000, Australia
* Correspondence: aonghus.mcnabola@rmit.edu.au

Abstract

The increasing deployment of pumps-as-turbines in small-scale hydropower applications in off-design conditions strengthens the need for the monitoring of the operation and maintenance (O&M) needs. PATs (pumps-as-turbines, pumps operated in reverse to generate electric current) are increasingly used because of their low cost as micro-hydropower plants; however, limited research has focused on their maintenance needs during operation. This is an important consideration given their use under conditions for which they were not originally designed. One of the most challenging O&M issues in hydromachinery is cavitation, which can harm turbines and reduce their efficiency. In this study, Computational Fluid Dynamics (CFD) was used for 15 different simulations of PAT configurations and their cavitation behaviour was investigated under varying inlet pressure and mass flow conditions. A cavitation strength indicator was developed using linear regression, describing the strength of cavitation from 0 (no cavitation) to 100 (extreme cavitation). This parameter depends on mass flow rate and head, which are easily measured parameters using standard sensors. With this approach, it is possible to monitor cavitation status in a continuous manner in a working PAT without the need for complex sensors. With this application, it is also possible to avoid costly damage, shutting down turbines when cavitation strength is exceptionally high.

Keywords: pumps-as-turbine; cavitation; CFD; hydropower; condition monitoring

1. Introduction

The use of pumps running in reverse as turbines—referred to as pump-as-turbines (PATs)—has become an increasingly attractive option for small-scale energy generation. One of the major advantages of PATs is their lower investment cost compared to traditional hydraulic turbines and easier maintenance [1]. Since they are usually standardized, mass-produced centrifugal pumps, replacement parts are readily available and cheap, compared to replacement parts for other types of hydro turbines [2]. Furthermore, the wide accessibility of these pumps improves their versatility as turbines and makes spare parts easier to obtain compared to traditional turbines.

In recent years, the growing interest in renewable energy solutions has pushed for more adaptable, flexible technologies, allowing for the optimization of energy generation in diverse conditions. For PATs, these systems might be deployed in either remote areas or more urbanized environments with frequent changes in energy demand. This flexibility is one of the primary reasons for their increasing adoption, as it enables energy generation in situations where traditional turbines might not be viable due to cost, space limitations, or regulatory constraints [1,3].

Like other types of renewable energy technologies, there is an increasing demand for hydro turbines to be adjusted to changing flow conditions. This requirement causes PATs to operate in conditions that can often be far from the best-efficiency point (BEP). While some aspects of cavitation in PAT systems have been studied—including numerical investigations such as those by Wenguang and Zhang (2017) [4]—the broader topic of PAT failure mechanisms remains insufficiently explored. As noted by Stephen et al. (2024) [5], there is a clear gap in the literature regarding the systematic analysis of failures in PATs, including cavitation-related damage. This highlights the need for further research focused not only on cavitation itself, but also on its role in overall system degradation and failure. Cavitation occurs when certain criteria are met, i.e., the local pressure in a fluid falls below its vapour pressure, causing the creation of vapour. It appears in low-pressure zones and forms bubbles, which collapse violently when they enter high-pressure areas [6]. This causes a sudden implosion, which results in intense shockwaves that can cause significant damage to metal surfaces, leading to the creation of pits and cavities [6,7]. On the other hand, cavitation also directly lowers the efficiency of working PATs, decreasing the amount of generated energy [8].

Additionally, cavitation negatively impacts the economic viability of PATs, especially when they are subjected to extended operation under cavitation. It leads to erosion of turbine components, reducing their operational lifetime and increasing maintenance costs [8]. In addition to wear, cavitation also generates excessive vibrations and noise [5]. For small-scale hydropower applications, where downtime and repair costs can quickly outweigh the benefits, understanding and mitigating cavitation becomes crucial. This is especially more pressing in recent times due to the aforementioned need for more flexible operation of hydropower devices [9].

Typical hydro turbines usually operate in relatively steady flow conditions, which allows for the analysis and mitigation of the chances of cavitation occurring, using, for example, dimensionless parameters. For this purpose, one of the most often used indicators is the Thoma number (σ) [10]. Utilizing this, it is possible to choose the proper type of turbine for given conditions and adjust its design accordingly. However, this approach leaves a large gap in the understanding of flow properties inside the turbine, which are not within the scope of these methods. This is because dimensionless numbers like the Thoma number cannot provide any information about local variations in pressure or flow velocity within the turbine [11]. These analytical tools focus on global operating parameters and neglect the flow behaviour, which plays a critical role in phenomena such as cavitation or efficiency losses [12].

For this reason, to understand the flow inside a PAT working in off-design conditions and analyse how its properties are related to the existence and strength of cavitation, there is a need for the application of modelling methods, particularly Computational Fluid Dynamics (CFD) [13]. CFD is considered a trustworthy and reliable tool for fluid flow analysis in various engineering applications. It enables detailed flow investigation, especially in complex geometries and under turbulent, transient, or multi-phase conditions. The most meaningful advantage of CFD is in its ability to provide data on local pressure and velocity distributions, which are often inaccessible through experimental methods [3]. As a result, CFD has become an essential method for studying internal flow phenomena in hydraulic machines.

While the Thoma number is often used in the design of conventional turbines, its application to PATs is limited due to the dynamic nature of their operating conditions. PATs often experience fluctuating flow patterns and varying loads, which makes it difficult to predict cavitation risk. This highlights the need for more accurate methods that take into account the local conditions within the turbine. Therefore, relying on the Thoma number or similar dimensionless parameters for cavitation prediction in PATs is not sufficient, especially in systems operating in off-design conditions [7]. The Thoma number is usually applied during the design phase of a hydropower plant to ensure that cavitation will not occur under relatively steady operating conditions. It is not designed to quantify the intensity of cavitation once it appears. In the case of PATs, they often operate under unstable flow conditions, so cavitation may still develop even if the model of PAT was selected based on the Thoma number. When cavitation occurs, the Thoma number offers a limited description of its severity. Therefore, additional indicators are needed to quantify cavitation intensity in such conditions.

Using CFD, internal flow properties can be linked to cavitation, evaluating the existence and strength of cavitation based on properties that are easily measurable in real time with digital sensors, like mass flow rate. The application of CFD also allows for examining cases that are hard to capture in real, working hydropower plants, either with extremely high or low flow rates or with high or low heads, which occur rarely during the year in natural conditions.

The goal of this work is to create a cavitation strength indicator, which will describe the strength of cavitation with a gradual scale, where 0 means no cavitation and 100 relates to an extreme case of cavitation. Its application will allow it to predict the cavitation in working PATs based on flow properties measured with digital sensors and let the user of the PAT know when the turbine should be shut off to protect its parts from damage. Additionally, CFD will be used as a tool to better understand the development of cavitation in the environment of increasing mass flow rate for a given PAT.

The key contribution of the paper is the analysis of cavitation in PATs. The cavitation in pumps has been extensively studied, but in reverse as a turbine, it has been neglected. This oversight is particularly important when considering that PATs operate in reverse under conditions they were not designed for, and as a result, PATs are more susceptible to cavitation. Understanding this failure mechanism in PATs is, therefore, crucial given their recent increased prevalence in hydropower practice, their increased susceptibility, and the increasing demand for flexible operation. Flexible operation of PATs is a particularly crucial consideration given that their typical efficient and safe operating envelope is much narrower than conventional turbines.

2. Methodology

2.1. The Goal of This Study

The goal of this study is to develop a CFD-based cavitation strength indicator (CI) that can be used to estimate the onset and severity of cavitation in a pump-as-turbine (PAT). To achieve this, a series of CFD simulations were performed for a selected PAT. The CFD model was validated against experimental results of the pump used as a PAT and experiencing cavitation. This experimental data is described in [5]. Simulation results were post-processed to extract flow parameters, which were then used to construct and evaluate the CI.

2.2. PAT Geometry

In this study, the model used was the KSB 050-32-200 centrifugal pump (manufacturer: KSB SE & Co. KGaA, Frankenthal, Germany), operated in PAT mode. The operating points for mass flow rate and head were chosen to cover three distinct regimes: pre-cavitation, cavitation inception, and fully developed cavitation. While the inception point is considered relatively fixed, the other two points were selected in an arbitrary but representative way to capture the general progression of cavitation severity.

Its specifications are summarized in Table 1. While the nominal impeller diameter is known from manufacturer data (170 mm), the exact geometry of the impeller blades—such as blade profile—was not available. These features, which significantly affect flow behaviour and energy conversion efficiency, were estimated using ANSYS software 2023 R1, especially Ansys Vista CPD 2023 R1, Ansys Bladegen 2023 R1. This process is more deeply described in Section 2.3.5.

Table 1. Specification of KSB 050-32-200 (pump mode).

Specification	Value
Brand/Model	KSB Etanorm 050-32-200
Max Pressure	<16 bar
Flow Rate	<604 L/min (<36 m^3/h)
Rotational Speed	960/1450/2900/3500 rpm
Power	0.75 kW
Temperature Range	−30 °C to +140 °C

The pump operates within a wide range of flow rates, which are detailed in the specifications. In PAT mode, the geometry of a chosen pump allows it to convert excess pressure into mechanical energy, thus acting as a turbine. The performance of PAT is dependent on its ability to maintain stable operation while handling varying flow conditions.

For accurate simulations, the corresponding data was used in CFD simulations to evaluate its performance in different operational scenarios.

Experimental Validation Data

The experimental dataset employed in this research originates from the test rig developed by Novara [14]. The test rig is schematically described in Figure 1 and consists of a 9.2 kW supply pump, a single-suction volute centrifugal pump used in a turbine mode (PAT) powered by an induction motor (5.5 kW). The pump used to construct the rig was a six-blade KSB ETN 050-32-200 (manufacturer: KSB SE & Co. KGaA, Frankenthal, Germany). To conduct measurements, two sensors were installed—a flow meter and a pressure meter (A, C). The inlet pressure is measured using a Gems Sensors Series 3000 relative pressure transducer with a 4–20 mA current output, offering an accuracy of ±0.25%. Flow is measured by an Omega FDT506 inline ultrasonic flow meter (Omega Engineering, Norwalk, CO, USA), also featuring a 4–20 mA current output, with a typical accuracy of ±2% and repeatability of ±0.2%. Downstream of the PAT, a Gems Sensors 3500 Series absolute pressure sensor (Gems Sensors, Plainville, MA, USA) (B) with a 4–20 mA current output is mounted, while two torque transducers (D and E) are positioned at the connection point between the PAT and the generator to monitor torque on the generator shaft and the PAT output shaft, respectively. At point D, Torque meter Datum Electronics, model M425-1D (Series 425) (Datum Electronics Ltd, East Cowes, UK). This offered a nonlinearity better than ±0.1% Full Scale and repeatability within ±0.05% Full Scale. Based on the manufacturer's specifications, the overall uncertainty of the instrumentation is estimated to be within ±0.5% for pressure, ±2% for flow, and ±0.1% for torque.

In the piping loop, a transparent pipe fragment is installed after the PAT, which allows the onset of cavitation to be observed (see Figure 2). All the measurements are carried out in the situation where PAT's RPM is equal to 1500. Behind the pump, the throttle valve is installed, which allows it to modify the flow rate. Using all of this data, a Thoma number was calculated for each case selected for numerical replication and for each of these cases, the state of cavitation was visually observed. Finally, the outcome of these measurements includes head, mass flow rate, Thoma number, and visual observation of cavitation state. Terms "intense cavitation", "cavitation", and "no visible cavitation" refer to purely visual observations made during the experiment. "Cavitation inception" denotes the moment when a vortex rope begins to form and becomes visible in the transparent section of the pipe behind the PAT. "No visible cavitation" corresponds to the phase before any visible vapour structures appear, while "cavitation" and "intense cavitation" describe the development or presence of a fully developed vortex rope. Experimental data is described in Table 2.

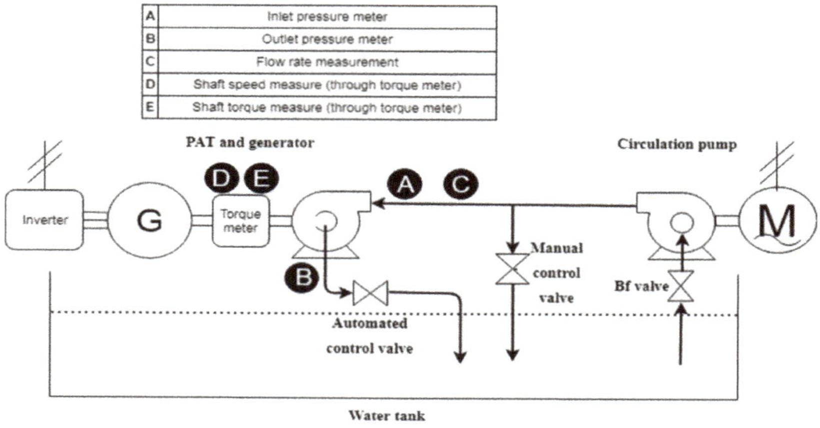

Figure 1. Experimental test rig (Novara, 2020) [14].

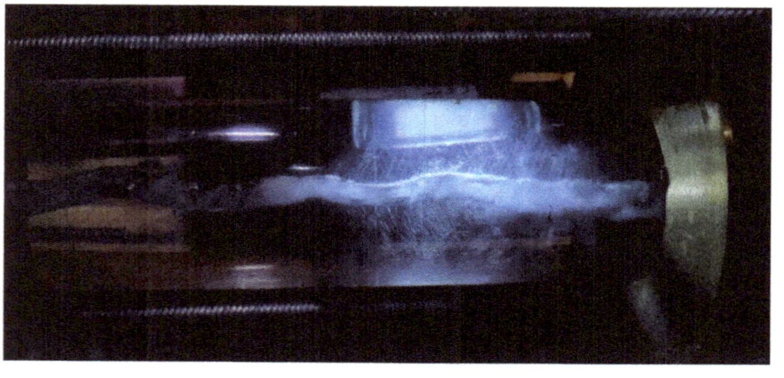

Figure 2. Cavitation occurring in the PAT test rig.

Table 2. Experimental data used for this research.

Case Number	Mass Flow Rate [kg/s]	Head [bar]	Thoma Number [-]	Cavitation State
1	2.189180392	46.55656055	0.0617	intense cavitation
2	2.275925362	46.92028188	0.062	intense cavitation
3	2.488844834	47.50478544	0.0631	cavitation
4	2.733307932	46.63453287	0.0633	cavitation
5	2.930455591	46.89430884	0.064	cavitation
6	3.127603251	46.63450627	0.0637	cavitation inception
7	3.427267693	47.02418544	0.065	cavitation inception

Table 2. *Cont.*

Case Number	Mass Flow Rate [kg/s]	Head [bar]	Thoma Number [-]	Cavitation State
8	3.797905293	46.9332874	0.0654	no visible cavitation
9	3.939851608	45.51738444	0.0656	no visible cavitation
10	4.129113361	45.23159742	0.0661	no visible cavitation
11	4.32623432	45.77723423	0.066	no visible cavitation
12	4.602267744	46.34871925	0.0663	no visible cavitation
13	4.941361718	43.49089469	0.0668	no visible cavitation
14	5.375086569	45.53026456	0.0676	no visible cavitation
15	5.564348322	44.37421357	0.0681	no visible cavitation

2.3. CFD Methodology

This section describes the numerical simulation used for modelling the PAT with application of CFD using Ansys CFX 2023 R1 software. The model was developed to analyze the properties of the flow and performance of the PAT. Simulations were carried out in steady-state "Turbo" mode, which is specifically created for turbomachinery and simplifies the setup for rotating domains. Later, these simulations were used as a starting point for transient simulations with the timestep 3.43×10^{-4} s and total simulation time 0.4 s, which corresponds to 10 full rotations of the impeller.

2.3.1. Governing Equations

The simulation was considered incompressible and Newtonian, using Reynolds-Averaged Navier–Stokes (RANS) equations. These describe the conservation of mass and momentum and depending on the chosen turbulence model, the conservation of other variables. As the k–ω turbulence model was implemented, these variables were k-turbulent kinetic energy and ω, the specific dissipation rate.

2.3.2. Turbulence Model

The k–ω SST (Shear Stress Transport) model is considered the industry standard for turbomachinery turbulence modeling [15,16] because it connects the advantages of k–ω and k–ε approaches, allowing for efficient and robust modelling of both near-wall and free-stream flow. For the inner layer, k–ω is applied and the algorithm switches to the k–ε approach for areas far from the wall, which can be considered as free-stream. This allows for good prediction of flow separation and adverse pressure gradients [15,16]. Additionally, this approach is suggested as the default approach and a robust methodology for turbomachinery applications in Ansys CFX documentation [17].

More advanced models, such as Large Eddy Simulation (LES) or Detached Eddy Simulation (DES), can be a source of more accurate description of cavitating flows. However, their application usually requires large meshes, which result in increased computational cost. For this study, the SST model was used as a practical balance between accuracy and computational feasibility, and a similar approach performed well in other cavitation-related simulations, including recent studies such as Khan et al. (2021) [18].

2.3.3. Rayleigh–Plesset Model

The Rayleigh–Plesset equation is used to describe the dynamics of one spherical vapour bubble in an incompressible environment. It describes the bubble inertia, liquid viscosity and surface tension. The Rayleigh–Plesset approach was proven in terms of accuracy and mass transfer between the liquid and vapour phase [19]. In general form, the Rayleigh–Plesset (RP) model was presented as Equation (1).

$$\rho_L\left(RR^{\cdot\cdot}+\frac{3}{2}R^{\cdot 2}\right)=P_\infty(t)-P_B(t)-\frac{2\sigma}{R}-4\mu_L\frac{R^{\cdot}}{R} \quad (1)$$

where *R(t)*—bubble radius as a function of time [m], Ṙ, R¨—first and second time derivatives of the radius (radial velocity and acceleration) [m/s, m/s^2], ρ_L—liquid density, assumed constant [kg/m^3], $P_\infty(t)$—ambient pressure far from the bubble [Pa], $P_B(t)$—pressure inside the bubble, assumed uniform [Pa], σ—surface tension at the liquid–vapour interface [N/m], μ_L—dynamic viscosity of the liquid [Pa·s]. In cavitation modelling, Rayleigh–Plesset model is considered the default and robust way of simulation [20].

2.3.4. Boundary Conditions

Boundary conditions were chosen to match those of the experimental data used for model validation [5]. Mass flow rate on the inlet was measured and the pressure was measured on the outlet and inlet during the experiments conducted by Stephen et al. [5] on the selected PAT.

Using Ansys CFX, it is—to apply the boundary template "Mass Flow inlet P-Static outlet". This template is suggested for turbines as it provides numerical stability and excels in calculating velocity and pressure profile of the inside of hydro machinery [21]. For walls, the no-slip boundary condition was used, assuming the velocity at walls of the turbine equal to zero, which is a fundamental assumption in fluid dynamics [22], except for foams. Boundary conditions are summarised in Table 3.

Table 3. Set-up used for CFD simulation.

Feature	Settings
Turbulence model	k-omega SST
Boundary condition	Inlet—fixed mass flow rate (2–6 kg/s) Outlet—average static pressure (0 Pa gauge pressure) Walls—No slip condition
Mass flow rate	2–6 kg/s
Cavitation model	Rayleigh–Plesset
RPM	1500
Number of cells	1.12 mln

At the inlet, a "Mass Flow Inlet" boundary condition was applied, where the total mass flow rate entering the domain was explicitly prescribed. Values used for simulations are shown in Table 4. The flow angle was set normal to the inlet surface. At the outlet, a Static Pressure Outlet boundary condition was applied. This combination of boundary conditions is common for turbomachinery and was used in previous studies, for example by Laouari, Ghenaiet (2016) [23] and Ortiz [24] (2019).

Table 4. Properties used in Vista CPD.

Parameter	Value
Volumetric flow rate	14.25 m^3/h
Total head	13.75 m
Rotational speed	1500 RPM
Inlet flow angle to impeller	Rayleigh–Plesset
Volute axial length	1500

2.3.5. Creation of Mesh

As the exact 3D geometry of the turbine's impeller and volute was not provided, it was necessary to obtain a similar geometry using reverse-engineering software. The initial step of its creation was to generate a 2D sketch of the impeller and volute, which will later be expanded into the 3rd dimension. Using documentation from the pump described in Table 1, several working conditions were used (listed in Table 4).

Based on given properties, Vista CPD was used to compute velocity triangles and optimize energy losses caused by the geometry to create 2D sketches. This is carried out using Equation (2):

$$\Delta h = \frac{u_2 c_2 - u_1 c_1}{g} \quad (2)$$

where Δh is a pressure head, which is equal to the product of head and gravitational acceleration g. u_1 and u_2 are the blade tangential speeds at inlet and outlet [m/s], c_1 and c_2 are the tangential components of the absolute velocity at inlet and outlet [m/s], g is the gravitational acceleration [m/s^2]. Tangential speeds are calculated using Equation (3).

$$u = \omega r = \frac{2\pi n}{60} * r \quad (3)$$

where u is the blade peripheral speed at radius r [m/s], ω is the angular velocity [rad/s], n is the rotational speed [rpm], r is the radial position from the shaft axis [m] [25]. Finally, the obtained sketches were calculated and drawn, as shown in Figure 3.

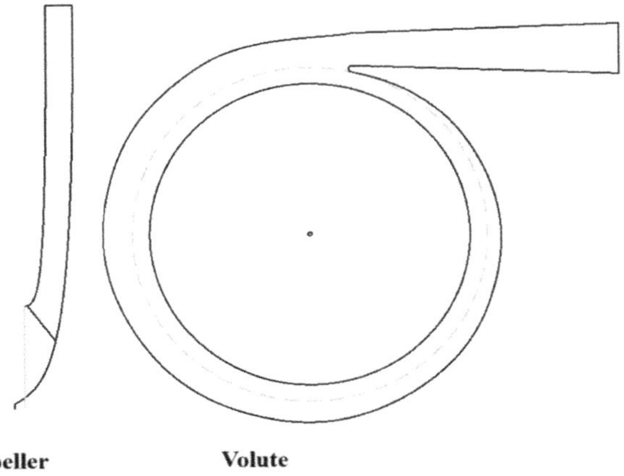

Figure 3. Two-dimensional sketches of the volute and impeller obtained through Vista CPD.

In the next step, BladeGen was used to find relationships to determine the impeller shape and positioning with an iterative approach. The aim of this optimization was to ensure that the created 3D impeller model achieves energy transfer while also maintaining high hydraulic efficiency, and the ability to be manufactured. Initially, the outlet blade angle β2 was found using the Euler equation. It was later used to calculate the relative velocity at the outlet, which can be computed with Equation (4).

$$w2 = u2 - c2 \quad (4)$$

Later, these relative velocities are applied to compute the tangent of β2 using Equation (5).

$$tan\ (\beta 2) = \frac{Q}{w2\ *\ Am} \tag{5}$$

where Q is the volumetric flow rate [m^3/s], Am is the meridional flow area at the blade section [m^2]. Once $\beta 2$ was computed, BladeGEN constructed a mean line, which is a curve that connects the midpoints between the pressure and suction sides of the blade profile. Its proper selection minimizes separation of the flow and turbulence, so the acceleration of flow is smooth across the blade length. Consequently, bladeGEN created a blade thickness distribution. The stacking line was created with axial stacking, which is a common choice for turbomachinery applications. Subsequently, the blade thickness distribution was applied by offsetting the camber line on pressure and suction surfaces according to a chosen profile shape. The thickness distribution has a direct impact on local pressure gradients and potential flow separation [25].

Figure 4 shows a distribution of Theta and Beta in degrees over the blade profile, which is displayed as the normalized chord length from the leading edge (LE) to the trailing edge (TE). The β curve represents the blade metal angle, while the θ curve shows the blade twist angle, derived from the relative rotation of each profile along the stacking axis. According to the BladeGEN guide, these angles can be described as:

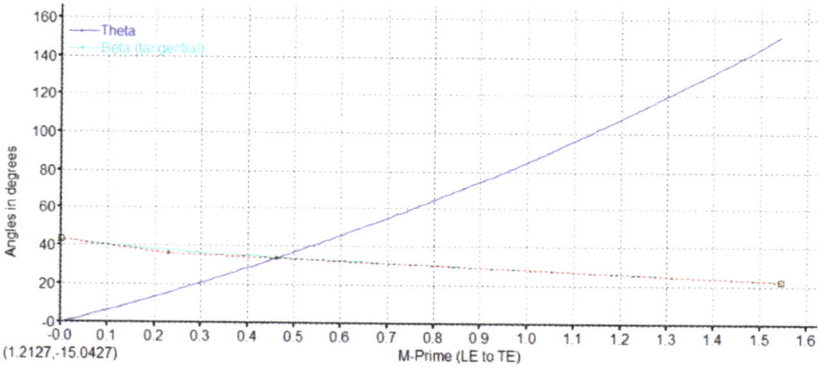

Figure 4. Distribution of angles over the blade profile.

β (Beta)—the blade metal angle, defined as the inclination of the blade surface relative to the local meridional direction. It reflects how the blade redirects the relative flow from inlet to outlet. The angle is typically measured at various spanwise sections (e.g., hub, mid-span, shroud) and varies from the leading edge (LE) to the trailing edge (TE).

θ (Theta)—the twist angle, which quantifies the rotation of the blade profile about the stacking axis. It describes how the orientation of each 2D profile changes along the span (from hub to shroud), affecting the alignment of the blade with the three-dimensional flow field [25].

In the next step, the meridional plane was generated from an outlet to the inlet (including the shroud and its contours), and the number of blades was set as 6 to reflect the existing PAT geometry and satisfy Equation (6).

$$\sigma = \frac{Z\ *\ Sm}{2\pi Rm} \tag{6}$$

where is the dimensionless blade solidity, Sm is the blade spacing and Rm is the mid-span radius [m] [25]. The result of this approach can be seen on Figure 5.

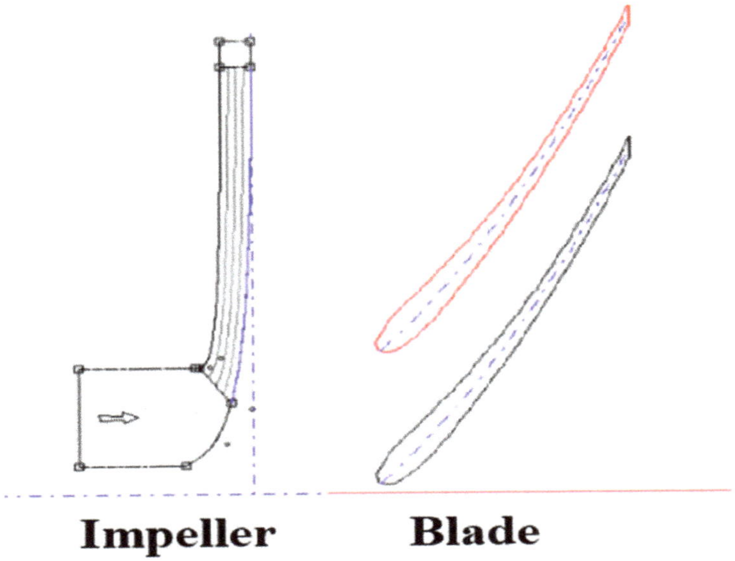

Figure 5. Geometry of 1/6 of impeller and 2D sketch of PAT's blade.

Using these parameters, BladeGEN generated the 3D shape of impeller, which consisted of the geometry of 1/6 of the impeller and the blade itself, which can be seen in Figure 6.

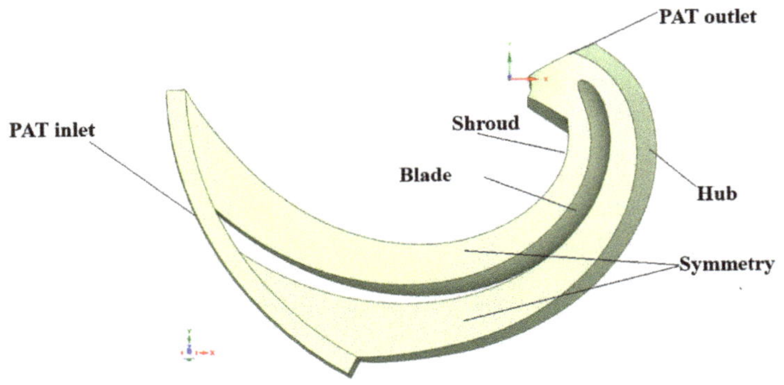

Figure 6. Three-dimensional geometry of 1/6 of the rotor.

Geometry created with BladeGEN (Figure 6) was later meshed, and the mesh consisted of three parts. The first part (rotor mesh) was a fully-structured mesh created with Ansys ICEM and the remaining two parts (the extension of an outlet, volute mesh) were created with the built-in ANSYS package meshing software. In ANSYS ICEM, the mesh was created from manually distributed blocks, and the distribution of blocks used is presented in Figure 7. Later, these meshes were connected with Frozen Rotor [21] interfaces and used as an input to the CFD simulation.

The rotor was created by revolution of the meshed geometry created in previous steps. It consisted of approximately 400×10^3 cells, while the volute was created with the element size of 2×10^{-3} m, with 10 boundary layers, achieving for both meshes the y+ value around 1, which is required for the application of k–ω SST turbulence modelling.

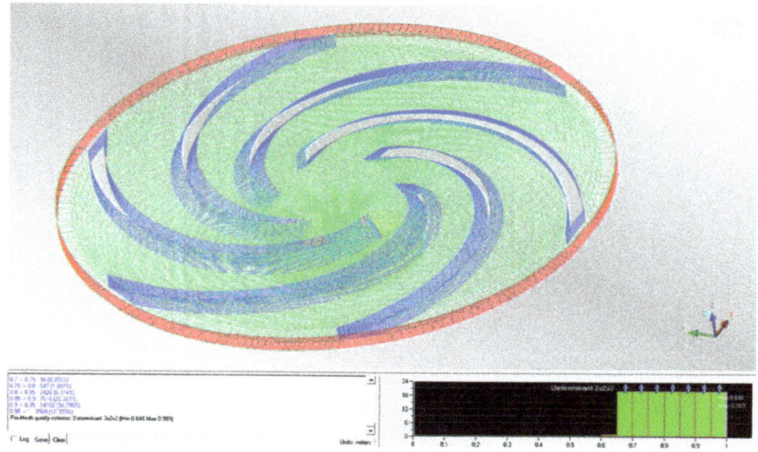

Figure 7. Block distribution of complete impeller after revolution.

2.3.6. Mesh Independence

To ensure mesh independence, 7 meshes were examined, consisting of 160 k, 195 k, 480 k, 640 k, 1120 k, 1400 k and 1550 k elements. As the experimental measurements included efficiency, this was examined using Equation (6), set with Ansys expression. Each mesh was created with identical boundary conditions and number of boundary layers, and they were examined on the same CFD setup (solver settings, SST k–ω turbulence model). The convergence criteria were to achieve the mean residual range of 10^{-5}, which became possible at meshes with the number of elements equal to 480 k and larger. The performance of each mesh was examined based on efficiency and its closeness to experimental results. The expected result of the mesh independence studies was that the turbine's efficiency should not fluctuate by more than 1% between successive mesh refinements, proving that the numerical solution is independent of the mesh size. The lowest size of mesh that could satisfy this requirement was the mesh size of 1.12 mln elements, which can be seen in Figure 8. The final mesh had skewness below 0.65 and orthogonality above 0.75, ensuring sufficient quality for cavitation simulations. The rotor region was meshed using a fully structured grid in Ansys ICEM, achieving skewness below 0.5 and orthogonality above 0.85, which improves accuracy in critical flow regions.

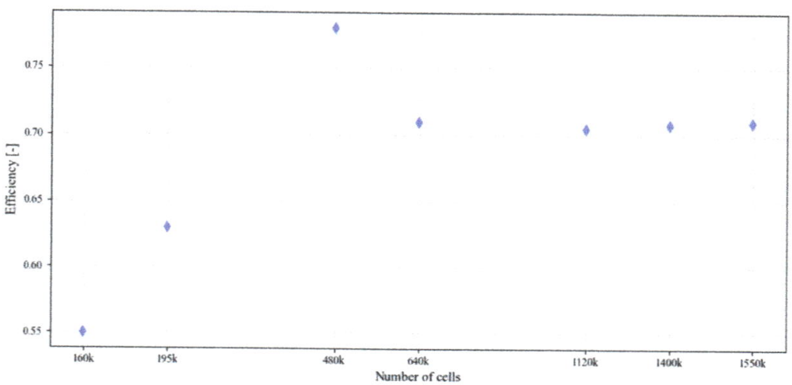

Figure 8. Mesh independence study.

2.3.7. CFD Setup

The CFD process was carried out using ANSYS CFX. To reflect the physical conditions of the working PAT, the starting conditions summarized in Table 2 were applied. Geometries with Vista CPD and BladeGen were created using the data listed in Table 4.

Because steady-state simulations failed to reproduce experimental results, the simulation was carried out in transient mode. The time step of 3.43×10^{-4} s was set to maintain a Courant number below 1, which is a critical value for accuracy.

The mean Courant number for the final mesh was 0.863, with a maximum of 1.5. Previous meshes had similar cell topology (they were only later refined), so it is assumed that their Courant numbers were also below 1. Furthermore, to improve the quality of the final solution, the rotor mesh was created fully structured in ICEM CFD with skewness around 0.5 and mean orthogonality 0.82, which allowed for larger cells near critical regions without compromising accuracy. The chosen time step allowed the solution to remain stable and consistent with physical observations. The rotational velocity was set to 1500 RPM and total simulation time was set to 0.4 s, which is approximately the time required for 10 full rotations of the rotor. This choice was made to allow for the flow stabilisation, ensuring sufficient time for flow development and later convergence of the simulation.

The discretization scheme used was a second-order upwind for all convective terms. For pressure, a central difference scheme was applied. Convergence was achieved at mean residual target of 10^{-5}. The simulation was set using CFX Turbo mode. Three components (impeller, volute, outlet extension pipe) were connected with a frozen rotor interface type, where only the impeller rotates with a speed of 1500 RPM. The rotation was negative to enable PAT mode. Finally, CFX recognized 7 boundaries, which were set as an inlet, outlet and wall Boundary type. For wall influence on flow, no-slip wall condition was set, representing null velocity of fluid at the wall, which is fundamental CFD assumption for liquids. In terms of working fluid, "water" from Ansys library was removed and replaced with the mix of liquid water and vapour water. To ensure that the liquid entering the computational domain consists only of liquid water, the mass fraction of liquid water on the inlet was set to 1 while the mass fraction of vapour water was set to zero. This condition was required to enable the cavitation simulation and change of phase during simulation. The chosen interphase transfer was the mixture-model, and mass transfer setting was set for cavitation, which was described with the Rayleigh–Plesset model. The interface length scale and mean diameter of the vapour bubble were set, respectively, to 1 mm and 2×10^{-6}, which are default values for this mode, while the saturation pressure was set at 2.34 kPa, which is the value of choice for water at a temperature of 293 K. As the temperature changes in the indoor location of the experimental measurement were negligible and the saturation pressure is not influenced by them in large quantities, this value was set as constant.

2.4. *Thoma Number Limitations and the Purpose of the New Indicator*

The Thoma number (σ) is a dimensionless parameter used in turbomachinery planning to determine the presence of cavitation. It is expressed as the ratio of the difference between the outlet pressure and the saturation pressure of the fluid to the hydraulic head, which is stated in Equation (7).

$$\sigma = \frac{Poutlet - Psat}{\rho g h} \tag{7}$$

where *Poutlet* is the static pressure at the turbine outlet [Pa], *Psat* is the saturation pressure of the working fluid [Pa], ρ is the density of fluid [kg/m^3], g is gravitational acceleration [m/s^2] and h is the total head [m].

Equation (7) aims to identify situations when local pressure falls below the vapour pressure which will lead to phase change, thus cavitation. When the Thoma number is below the critical threshold specific to the turbine and design conditions, cavitation is likely to occur. It works mainly as a binary indicator, either showing the cavitation presence or lack of it. To address this limitation, a cavitation index (later called cavitation strength) was developed, which is created using the value of vapour volume fraction achieved through CFD simulation. CI will provide a quantitative estimate of the cavitation intensity, describing its power on a normalized scale from 0 (no cavitation) to 100 (extremely intense cavitation). While the Thoma number is a valuable and robust preliminary metric for the design stage of turbines working in steady conditions, CI enables deeper insight into the scale of this phenomenon. Additionally, when created, it can be computed using non-sophisticated data, such as pressure on the inlet and mass flow rate, which can be derived from standard sensors.

After creation of the meshes, they were used for CFD simulations. The obtained results were measured for efficiency using Equation (8). To check the robustness of simulations, their results were compared with experimental results.

$$n = \frac{T * \omega}{\Delta P * V} \quad (8)$$

where T—Torque [N*m], ω—angular velocity [rad/s], ΔP—change of pressure between the inlet and the outlet [Pa], V—volumetric flow rate [m^3/s]

When the solution was considered mesh independent at 1.12 mln total cell number, an entire flow curve of PAT was simulated using CFD. Later, these results were compared with experimental results on a previously created test rig with obtained efficiencies of the turbine at different flow rates. The result of this comparison can be seen in Figure 9.

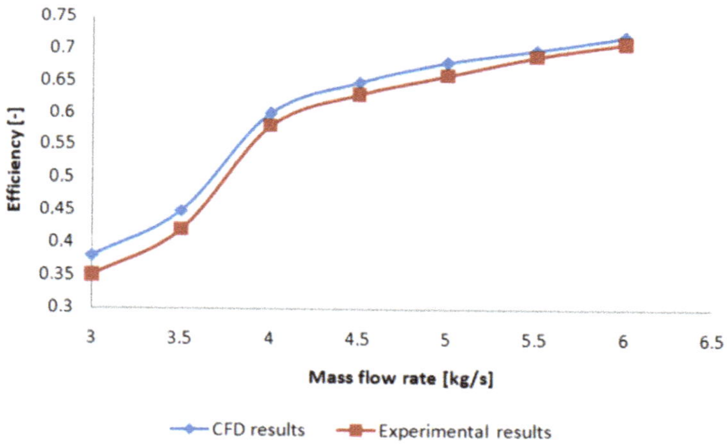

Figure 9. Efficiency comparison between experimental and CFD results.

When the CFD results were trustworthy enough to display proper efficiency of the turbine, cavitation prediction was added. Using the Rayleigh–Plesset model, it was decided to keep most of the properties at default for Ansys CFX turbomachinery application, while Saturation pressure was also set to default for the water at a temperature of 20 °C. According to experimental results, it could be seen that the cavitation was visible after the mass flow rate decreased to around 3.42 kg/s, which could also be seen in the CFD results (Figure 10). Further lowering the mass flow rate resulted in an increase in cavitation intensity in the experimental test rig.

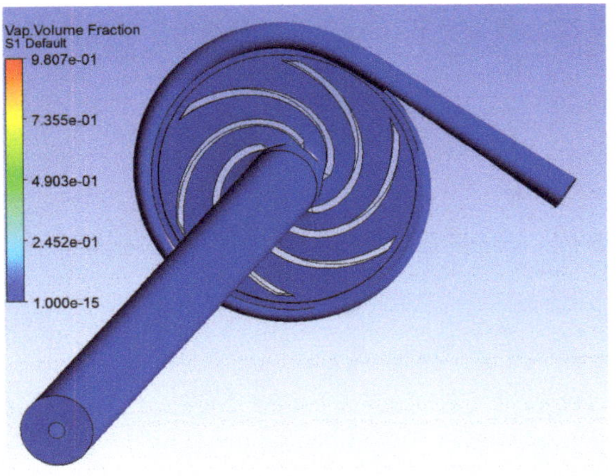

Figure 10. Vapour volume fraction at mass flow rate 4.6 kg/s—no cavitation.

2.5. Data Used for CI Indicator Construction

To quantify the cavitation strength, vapour volume fractions were categorized into three classes: Cells75 (vapour volume fraction > 0.75), Cells50 (0.50 < vapour volume fraction ≤ 0.75), and Cells25 (0.25 < vapour volume fraction ≤ 0.50). This classification allows weighting of the most intense cavitation zones more heavily. The number of cells with a vapour volume fraction in these brackets was considered. These quantities were calculated relative to the total number of cells in the mesh. Subsequently, an objective function was created to determine the cavitation intensity on a scale from 0 to 100. These values were then linked to the head and mass flow rate to predict cavitation.

2.6. CI Development

The algorithm utilizes two variables-head and mass flow rate to determine the strength of cavitation ongoing inside the PAT. Mass flow rate is a boundary condition that is set during the simulation. To calculate the cavitation strength, data from CFD was post-processed to obtain the percentage of cells with values of vapour volume fraction higher than 0.75, 0.5 and 0.25, respectively. All of these parameters were obtained directly from CFD analysis. For every case, the value of cavitation strength was calculated using Equation (9).

$$Cavitation\ strength = Cells75 * 3 + Cells50 * 2 + Cells25 \tag{9}$$

where $Cells75$, $Cells50$ and $Cells25$ are the percentage of cells that show the vapour volume fraction over, respectively, 75%, 50% and 25%. Cells with vapour volume fraction lower than 25% were ignored in this approach.

Later, results were divided by the highest-obtained value of cavitation strength to normalize them. In the next step, all were multiplied by 100 to achieve results between 0 and 100, where 0 means no cavitation and 100 means the most extreme cavitation out of the examined group. This improves interpretability by scaling results relative to the worst observed condition (cavitation strength 100), which makes it easier to understand (cavitation strength becomes the description of cavitation severity compared to the worst scenario examined with CFD). The results obtained for 15 cases examined in this research are presented in Table 5. The amount was chosen to capture the variation in flow across the experimental range measured.

Table 5. Results.

Case Number	Mass Flow Rate [kg/s]	Head [bar]	Thoma Number [-]	Cavitation Power [-]	Cavitation State
1	2.189180392	46.55656055	0.0617	81.16	intense cavitation
2	2.275925362	46.92028188	0.062	74.85	intense cavitation
3	2.488844834	47.50478544	0.0631	61.26	cavitation
4	2.733307932	46.63453287	0.0633	56.8	cavitation
5	2.930455591	46.89430884	0.064	53.46	cavitation
6	3.127603251	46.63450627	0.0637	46.41	cavitation inception
7	3.427267693	47.02418544	0.065 (critical)	47.40	cavitation inception
8	3.797905293	46.9332874	0.0654	33.65	no visible cavitation
9	3.939851608	45.51738444	0.0656	29.52	no visible cavitation
10	4.129113361	45.23159742	0.0661	25.30	no visible cavitation
11	4.32623432	45.77723423	0.066	16.81	no visible cavitation
12	4.602267744	46.34871925	0.0663	0.00410	no visible cavitation
13	4.941361718	43.49089469	0.0668	0.000108	no visible cavitation
14	5.375086569	45.53026456	0.0676	0.000068	no visible cavitation
15	5.564348322	44.37421357	0.0681	0.00040	no visible cavitation

To construct the algorithm, head and mass flow rates were related to cavitation strength using linear regression.

3. Results

The first part of the CFD research was focused on matching the CFD efficiency results with the experimental measurements of Stephen et al. [5].

The simulation results showed a very good agreement with the experimental results across the examined mass flow rates. The coefficient of determination R^2 equals 0.973, indicating that the created CFD model is able to accurately reflect the physical experiments. The root mean square error RMSE was equal to 2.14%, which is a further confirmation that the differences are low.

In general, lowering the parameter of mass flow rate as a boundary condition increases the intensity of cavitation, with the inception starting around 3.42 kg/s. It is visible that cavitation starts around the shroud side of the blade and extends to the outlet, creating a cavitation rope. With a further decrease in mass flow rate at the inlet, the number of mesh cells with a vapour volume fraction higher than 0.75, 0.5 and 0.25 increases. It is necessary to mention that the volute is generally prone to cavitation and the most susceptible parts are the impeller blades. The results of the simulation agree with the observational results; however, it is difficult to accurately quantify the cavitation based on visual perception. Using the created test rig, cavitation can be observed only downstream of the PAT (around points D and E in Figure 1), which leaves no information about the cavitation inside the turbine, where blades are susceptible to cavitation. Based on previous research by Demirel [26], it can be seen that cavitation onset takes place inside the turbine. While the cavitation rope is a visible part seen behind the rotor, there is also cavitation taking place on the blades, especially on the shroud side (Figure 11).

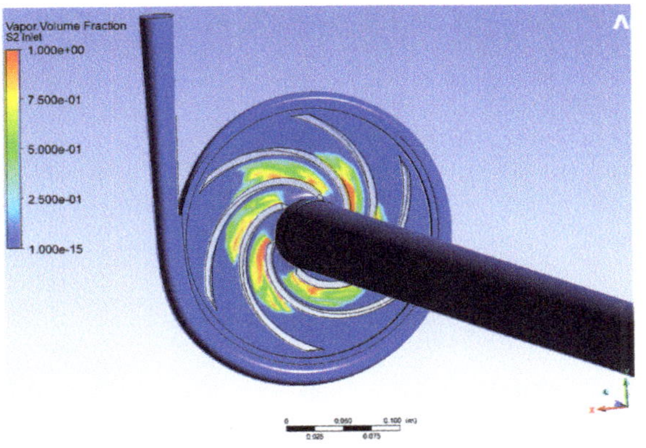

Figure 11. Vapour volume fraction at mass flow rate at 2.7 kg/s—intense cavitation.

3.1. Prediction Results

It can be seen that the Thoma number is described with nearly linear distribution, where lowering the mass flow rate—the most important factor for cavitation strength—significantly affects it.

The obtained empirical relationships between head and mass flow rate are displayed in Equation (10).

$$Cavitation\ power = -2.1361 + (-20.7194) * mass\ flow + (2.4884) * head \quad (10)$$

Equation (10) was obtained using a linear regression technique implemented in Python 3.10, using scientific libraries NumPy, pandas and scikit-learn. The model received a score of R^2 of 0.86, indicating that there is a significant degree of correlation between the predicted and received cavitation power values. Additionally, the RMSE at 6.65 clarifies that the model maintains a reasonable prediction accuracy across the given dataset. Interpreting the cavitation behaviour, it can be seen that the mass flow rate has a negative coefficient, which agrees with expectations—increasing the flow rate influences local velocities, lowering static pressures, which stimulates cavitation. The head has a negative coefficient, which means it also stimulates cavitation onset.

It can be seen that the Thoma number shows relatively small fluctuations as the mass flow rate decreases, even when it is the primary factor influencing cavitation power. In contrast, cavitation power shows significantly greater sensitivity, indicating that cavitation develops in a nonlinear manner. Unlike the Thoma number, which decreases slowly and near linearly, cavitation power provides a more dynamic and informative representation of the intensity of cavitation. This makes it a good indicator for assessing how vapour formation evolves under changing flow conditions. The stronger response of cavitation power to reductions in mass flow rate suggests that it captures the physics of cavitation more directly, making it a valuable tool in characterizing cavitation behaviour beyond what the Thoma number alone can show. This is especially visible in lower mass flow rates, where the increase in cavitation power between cases varies significantly, while Thoma number values present similar patterns. According to Figure 12, it can be seen how the cavitation power value depends on CFD cases, where the cavitation power value is computed for points where the simulation was carried out. It is visible that cavitation is a nonlinear phenomenon and while it shows rather simple dependency (lower mass flow rate stimulates cavitation and higher head stimulates cavitation), it can be seen that the points are not linearly displaced. In Figure 13, the results of Equation (10) are presented on a heatmap, showing that the predictor indicates a similar trend that was visible for higher mass flow rates.

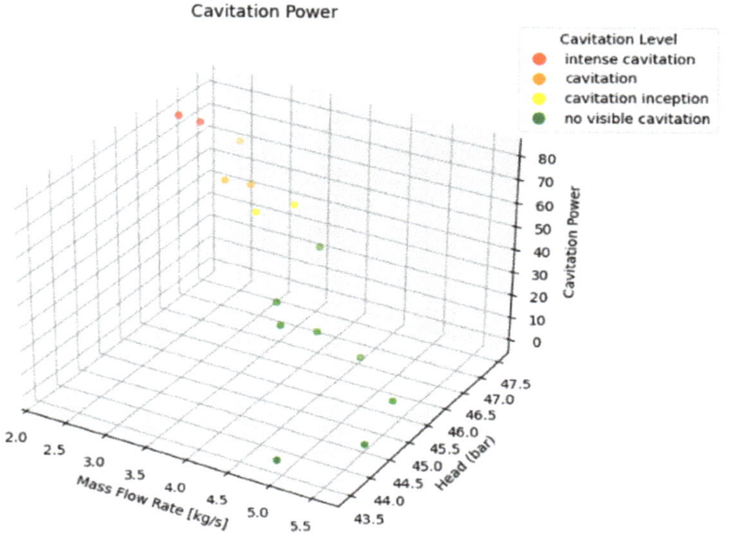

Figure 12. Cavitation power values of measured CFD setups.

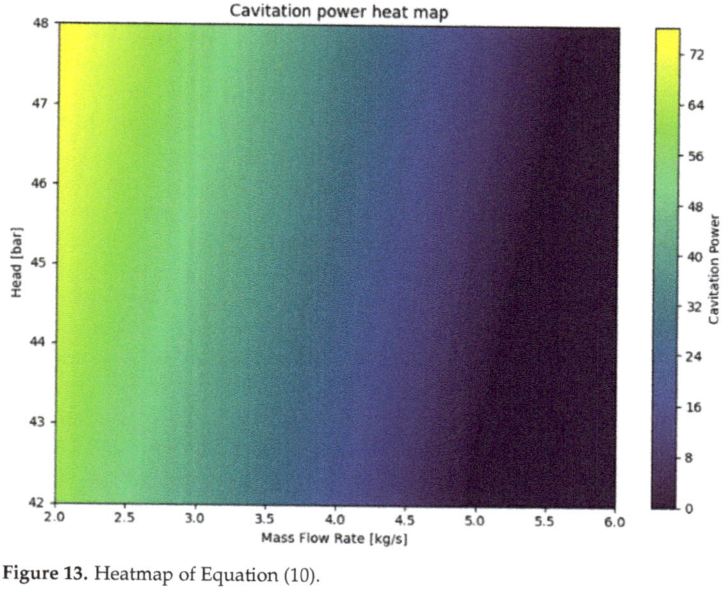

Figure 13. Heatmap of Equation (10).

Figure 14 displays what is the percentage difference between subsequent Thoma numbers of adjacent cases and how it relates to the percentage states of cavitation power. The first point compares the difference between states 1 and 2, and, respectively, point 13 compares states 13 and 14. The last point was removed with a value over 488%. This is caused by the division of very small numbers, which relate to a lack of cavitation and can be considered close to zero.

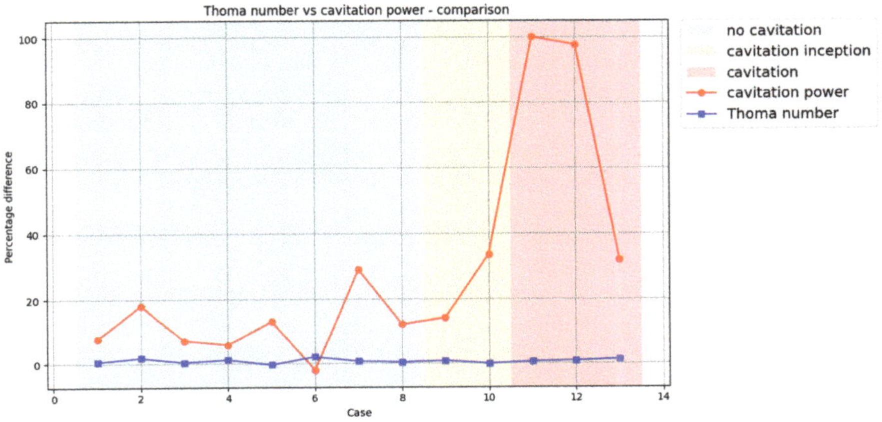

Figure 14. Comparison of Thoma number and cavitation power.

3.2. Sensitivity Studies

A sensitivity analysis was performed at five representative operating points (cases 1–5 in Table 2) to assess how the CI metric responds to potential measurement noise in head and flow inputs. Variability levels of 0.5%, 1%, and 5% were applied. The results show that CI remains relatively stable under low input noise ($\leq$1%), but larger deviations (5%) can significantly affect CI. The indicator was found to be more sensitive to head than to mass flow. Figure 15 summarizes the results, including cases where both inputs were perturbed simultaneously ("2 var in Figure 15").

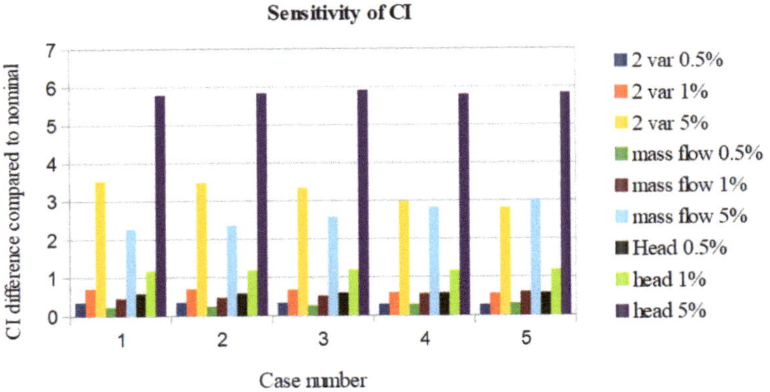

Figure 15. Sensitivity studies.

4. Discussion

While the value of R^2 equal to 0.86 in Equation (10) suggests that the prediction method is solid, the value of RMSE at 6.65 additionally confirms it. The purpose of this approach is to initially determine the strength of cavitation using simple input data, which is fulfilled. To address the question of how significant the error is, it is necessary to consider the application of real-time prediction with this algorithm. According to Table 5, cavitation starts with a 3.42 kg/s mass flow and a cavitation strength equal to 47.4. Similarly to the Thoma number, there is no hard threshold for the inception of cavitation and this number will always be dependent on the case and chosen setups. While adjusted to the specific case, this approach should always keep as much space as possible for states where cavitation is present, which means, the lower the cavitation inception is (in terms of cavitation power value), the better the accuracy of this prediction. It is possible that further expansion of the data (more cases with different operating conditions) and more specific curve-fitting strategies might address this problem.

The developed approach can serve as a solid basis for further validation and development. Other regression models could be applied (especially polynomial regression with different levels of polynomials). Cavitation can be predicted with CFD and its results can be considered as input data for the building of predictive indicators. The data that can be acquired with CFD is abundant and it is possible that many parameters can be used for the creation of likewise indicators. However, it is important to choose easily measurable properties of flow that can be read with the simplest possible sensors (due to economic reasons). This work only uses head and mass flow rate as they are easily measurable; however, it is very likely that this kind of indicator can be expanded with other properties of flow, allowing for more accurate prediction. Additionally, it is possible that a larger pool of simulation results would allow further adjustment of the prediction ability of this approach. Furthermore, the work does not address the question of whether the approach can be generalized for other geometries (only one PAT geometry was used in the research) and could produce an error in prediction.

The main reason for the application of CFD in this research is to find a method of accurate quantitative description of cavitation. Visual observation of cavitation intensity is highly subjective and prone to significant error, making it unreliable. In case of the existence of other methods that could very accurately describe the cavitation intensity in an experimental environment, this approach could be repeated with a significant increase in the amount of data.

As presented in Figure 14, the Thoma number exhibits a relatively small and nearly linear variation across the operating points, typically within a $\pm 5\%$ range. Visual observations of cavitation inception and development show that the cavitation progresses in a nonlinear manner. There is an initial region where variations in the Thoma number have no impact on the cavitation (non-cavitating regime), followed by a narrow transition zone in which cavitation initiates abruptly (cavitation inception), and finally, a regime of rapidly intensifying cavitation (fully developed cavitation). The Thoma number does not capture these transitions effectively, as it continues to increase uniformly across all regimes—regardless of whether cavitation is present or not. In contrast, CP remains stable in the non-cavitating range but exhibits a significant increase precisely at the onset of cavitation, followed by a second large rise corresponding to the transition into intense cavitation. These stepwise changes in CP are consistent with visual evidence of vapour formation, underscoring the indicator's ability to reflect critical changes in cavitation dynamics. This analysis shows that CI is a more sensitive and informative indicator of cavitation onset and severity than the Thoma number.

5. Conclusions

As CFD is a viable tool for the examination of cavitation occurring in the turbine, data provided by it can be used for building predictive models.

This study demonstrates that cavitation strength in a pump-as-turbine (PAT) can be reasonably predicted using simple and easily measurable parameters such as head and mass flow rate. In general, cavitation strength is able to predict cavitation like the Thoma number, but with improved accuracy in edge cases. The developed linear regression model achieved a high R^2 value, indicating good predictive potential. The simplicity of the method allows for quick, preliminary assessments of cavitation development, which can be particularly useful in real-time monitoring scenarios. While the prediction error is acceptable for distinguishing between strong and weak cavitation states, it might not capture more subtle behaviours. The findings using simple inputs can still offer a practical and scalable approach for early-stage diagnostics.

Furthermore, this work highlights the importance of using CFD simulations as a source of data for indicator development, particularly when visual methods fail to provide reasonable accuracy. The use of easily obtainable parameters increases the feasibility of implementation in real-world applications. To further improve the accuracy and generalizability, future development should consider expanding the dataset with more varied operating conditions and geometries, and exploring advanced regression techniques such as polynomial models or machine learning approaches.

Overall, the presented approach provides a solid foundation for developing practical cavitation prediction tools.

Author Contributions: Writing—original draft, M.J.; Writing—review & editing, A.M.; Supervision, A.M.; Project administration, A.M.; Funding acquisition, A.M. All authors have read and agreed to the published version of the manuscript.

Funding: This research was funded by the European Commission under Horizon Europe programme grant number 101122167 (iAMP-Hydro project). The APC was funded by Horizon Europe.

Data Availability Statement: The original contributions presented in the study are included in the article, further inquiries can be directed to the corresponding author.

Conflicts of Interest: The authors declare no conflicts of interest.

Abbreviations

The following abbreviations are used in this manuscript:

CFD	Computational fluid dynamics
PAT	Pump-as-Turbine
CI	Cavitation indicator

References

1. Kan, K.; Binama, M.; Chen, H.; Zheng, Y.; Zhou, D.; Su, W.; Muhirwa, A. Pump as turbine cavitation performance for both conventional and reverse operating modes: A review. *Renew. Sustain. Energy Rev.* **2022**, *168*, 112786. [CrossRef]
2. Williams, A. *Pumps as Turbines: A User's Guide*; Intermediate Technology Publications: Rugby, UK, 1995.
3. Plua, F.; Hidalgo, V.; Cando, E.; Pérez-Sánchez, M.; López-Jiménez, P. Pumps as Turbines (PATs) by Analysis with CFD Models. *Int. J. Adv. Sci. Eng. Inf. Technol.* **2022**, *12*, 1098–1104. [CrossRef]
4. Li, W.; Zhang, Y. Numerical Simulation of Cavitating Flow in a Centrifugal Pump as Turbine. *Proc. Inst. Mech. Eng. Part E J. Process Mech. Eng.* **2016**, *232*, 135–154. [CrossRef]
5. Stephen, C.; Basu, B.; McNabola, A. Detection of Cavitation in a Centrifugal Pump-as-Turbine Using Time-Domain-Based Analysis of Vibration Signals. *Energies* **2024**, *17*, 2598. [CrossRef]

6. Orlandi, F.; Montorsi, L.; Milani, M. Cavitation analysis through CFD in industrial pumps: A review. *Int. J. Thermofluids* **2023**, *20*, 100506. [CrossRef]
7. Lu, J.; Li, J.; Zhang, C.; Zhou, Y.; He, Y. Research on the Cavitation Characteristics of Pump Turbines Based on Mode Decomposition. *Processes* **2025**, *13*, 732. [CrossRef]
8. Al-Hashmi, S. *Detection and Diagnosis of Cavitation in Centrifugal Pumps*; The University of Manchester: Manchester, UK, 2005.
9. March, P. *Flexible Operation of Hydropower Plants*; Electric Power Research Institute (EPRI): Palo Alto, CA, USA, 2017. Available online: https://www.researchgate.net/publication/317789873_Flexible_Operation_of_Hydropower_Plants (accessed on 30 March 2025).
10. Manderla, M.; Kiniger, K.; Koutnik, J. Improved pump turbine transient behaviour prediction using a Thoma number-dependent hillchart model. *IOP Conf. Ser. Earth Environ. Sci.* **2014**, *22*, 032039. [CrossRef]
11. Shahzer, M.A.; Kim, S.-J.; Cho, Y.; Kim, J.-H. Effect of Thoma Number on Cavitation Inside a Francis Turbine. *AFORE* **2021**, *2021*, 419.
12. Cavitation. Available online: https://archive.nptel.ac.in/content/storage2/courses/112104117/chapter_7/7_12.html (accessed on 5 June 2025).
13. Singh, P.M.; Chen, C.; Chen, Z.; Choi, Y.-D. Cavitation Characteristics of a Pump-turbine Model by CFD Analysis. *KSFM J. Fluid Mach.* **2015**, *18*, 49–55. [CrossRef]
14. Novara, D. Hydropower Energy Recovery Systems: Development of Design Methodologies for Pump-as-Turbine in Water Networks. Ph.D. Thesis, Trinity College Dublin, Dublin, Ireland, 2020.
15. Wilcox, D.C. *Turbulence Modeling for CFD*, 2nd ed.; DCW Industries: Anaheim, CA, USA, 1998; p. 174.
16. SST k-Omega Model. Available online: https://www.cfd-online.com/Wiki/SST_k-omega_model (accessed on 2 March 2025).
17. ANSYS Inc. *ANSYS CFX User Guide, Release 2023 R1*; ANSYS Inc.: Canonsburg, PA, USA, 2023.
18. Khan, A.; Irfan, M.; Niazi, U.M.; Shah, I.; Legutko, S.; Rahman, S.; Alwadie, A.S.; Jalalah, M.; Glowacz, A.; Khan, M.K.A. Centrifugal Compressor Stall Control by the Application of Engineered Surface Roughness on Diffuser Shroud Using Numerical Simulations. *Materials* **2021**, *14*, 2033. [CrossRef] [PubMed]
19. Preston, A.; Colonius, T.; Brennen, C.E. Toward Efficient Computation of Heat and Mass Transfer Effects in the Continuum Model for Bubbly Cavitating Flows. Available online: https://caltechconf.library.caltech.edu/95/ (accessed on 5 June 2025).
20. ANSYS Inc. Cavitation Modeling in ANSYS CFX. Available online: https://ansyshelp.ansys.com/public/account/secured?returnurl=/Views/Secured/corp/v242/en/cfx_thry/cfxMultInteCavi.html (accessed on 13 March 2025).
21. ANSYS Inc. *ANSYS CFX Theory Guide, Release 2023 R1*; ANSYS Inc.: Canonsburg, PA, USA, 2023.
22. Zhu, Y.; Granick, S. Limits of the Hydrodynamic No-Slip Boundary Condition. *Phys. Rev. Lett.* **2002**, *88*, 106102. [CrossRef] [PubMed]
23. Laouari, A.; Ghenaiet, A. Simulations of steady cavitating flow in a small Francis turbine. *Int. J. Rotating Mach.* **2016**, *2016*, 4851454. [CrossRef]
24. Ortiz, J. CFD Modelling of Cavitation in Hollow Jet Dispersive Valves. In Proceedings of the 38th IAHR World Congress, Panama City, Panama, 1–6 September 2019.
25. ANSYS Inc. *ANSYS BladeGen User Guide, Release 2023 R1*; ANSYS Inc.: Canonsburg, PA, USA, 2023.
26. Demirel, G.; Ayli, E.; Celebioglu, K.; Taşcıoğlu, Y.; Aradağ Çelebioğl, S. Experimental determination of cavitation characteristics of hydraulic turbines. In Proceedings of the 2015 World Congress on Engineering, London, UK, 1–3 July 2015.

energies

MDPI

Article

Design and Development of Crossflow Turbine for Off-Grid Electrification

Asfafaw H. Tesfay [1,2,3,*], Sirak A. Weldemariam [2,3] and Kalekiristos G. Gebrelibanos [1,2,3]

1 Department of Civil and Environmental Engineering, Norwegian University of Science and Technology, S.P. Andersens Veg 5, 7491 Trondheim, Norway; kalekirstos.g.gebrelibanos@ntnu.no
2 Institute of Energy, Mekelle University, Mekelle P.O. Box 231, Ethiopia; sirak.aregawi@mu.edu.et
3 School of Mechanical and Industrial Engineering, Ethiopian Institute of Technology-Mekelle, Mekelle University, Mekelle P.O. Box 231, Ethiopia
* Correspondence: asfafaw.h.tesfay@ntnu.no or asfafaw.haileselassie@gmail.com

Abstract

Investing in large-scale hydropower is on the rise in Ethiopia in accordance with the country's climate-resilient green economy strategy. Rural electrification is a top priority on the development agenda of the country, with very limited off-grid interventions. Although small-scale hydropower can bring various social and economic benefits compared to other off-grid solutions, it is hardly localized in the country. The motivation for this research is to break this technological bottleneck by synergizing and strengthening the local capacity. Accordingly, this paper presents the full-scale crossflow turbine design and development process of a power plant constructed to give electricity access to about 450 households in a rural village called Amentila. Based on a site survey and the resource potential, the power plant was designed for a 125 kW peak at 0.3 m^3/s of discharge with a 53 m head. The crossflow was selected based on the head, discharge, and simplicity of development with the available local capacities. The detailed design of the turbine and its auxiliary components was developed and simulated using SolidWorks and CFD ANSYS CFX. The power plant has a run-of-river design, targeting provision of power during peak hours. This study demonstrates an off-grid engineering solution with applied research on the water–energy–food–environment nexus.

Keywords: small hydropower; crossflow turbine; rural electrification; off-grid electrification; run-of-river; renewable energy; hydropower development

1. Introduction

Hydropower is set to play a prominent role in many of the world's major electrical power grids in the 21st century. It also contributes almost a fifth of global electricity generation and more than half of global renewable electricity generation [1]. In this view, Ethiopia is often described as the water tower of northeastern Africa, with 45,000 MW of installable hydropower potential. In recent years, hydropower development has become the focus of attention in this country. Exploiting this resource would bring various social, environmental, and economic advantages to the country; most importantly, it would improve energy poverty and allow the country to strengthen its climate-resilient economic strategy. However, a lack of access to modern energy remains a chronic problem, especially in rural areas. Rural electrification through national grid extension is expensive and difficult, as more than 80% of the population lives in scattered rural village settings. On the other hand, the average annual potential to exploit water resources for off-grid applications through small-scale hydroelectric development is about 20 TWh/year [2]. This potential can contribute to the country's short–medium-term off-grid electrification strategy [3].

Like in other developing countries, rural electrification is a priority development agenda in Ethiopia. However, few appropriate off-grid solutions have been implemented to date [2]. In 2017, the country launched a "light to all" strategy to provide access to about seven million rural and deep rural households without national grid connectivity (about 35% of the population) and achieve universal access by 2025. To support this strategy, there was a plan to build more than eighty-five small and micro-hydropower plants. However, complete failure of this plan has occurred due to the implementation and unavailability of local technology.

Small and micro-hydropower plants are known for their ability to supply electricity in rural areas with environmentally sound, affordable, and adequate energy production. This would lead to increased employment opportunities, poverty alleviation, improvements in local living and cultural standards, environmental protection, and economic development in remote areas. In addition, major constraints associated with large hydro projects can usually be avoided. Moreover, small hydropower (SHP) schemes have short lifecycles and low investment costs per kW of installed capacity. It is also possible to size a system to meet a specific power demand based on the resource potential of a site and the available financing.

Ethiopia has a high installed capacity of standalone SHP plants. This capacity is estimated to be about 10% of the total potential (1500–3000 MW), though this might vary with seasonal rainfall and increased small-scale irrigation work [2]. The motivation for this paper is to demonstrate an ongoing project of small hydropower development from scratch using the local capacity with the objective to alleviate the lack of local technology that can be adopted, expanded, and scaled up by the government and community.

This paper presents findings from an applied research project on off-grid electrification, with a primary focus on optimized design and development of crossflow turbines, i.e., a novel approach within the Ethiopian context. This paper is organized in five sections. Section 1 provides an introduction and background information, and Section 2 details the employed methodology. The design of the crossflow turbine is presented in Section 3, followed by the results and discussion in Section 4. Section 5 covers the manufacturing process, and Section 6 provides conclusions and recommendations.

1.1. Site Description: Amentila Small Hydropower Plant

This paper presents only the turbine part of an SHP plant with a 125 kW generating capacity that is under construction for the rural community of Amentila Village. The power plant is found in the southeastern zone of Tigray (Northern Ethiopia), Enderta district, Amentila and Lihama villages and covers about 31.5 km^2 of catchment area. The village has a population of about 3000 that is fully dependent on agriculture. Around 450 households were considered to be electrified by the SHP technology in this phase. The power plant is a run-of-river (RoR) type, which depends directly on a natural flowing river and has a higher potential for a cascading project in phase two. Partial views of the village and project site are shown in Figure 1.

N.B.: The diversion was connected to a small tunnel measuring about 20 m by an open channel. The overall open channel between A and B was about 650 m long, and the penstock that connected B and C was 75 m long.

The available head and stream flow were used to estimate the power plant's capacity and design. Head measurements were taken at alternative locations using GPS around the entire station. Similarly, flow measurement was conducted using a current meter and the volumetric method during the driest season of the year (the worst-case scenario).

Figure 1. Amentila small hydro site: (**A**) diversion, (**B**) reservoir, and (**C**) powerhouse.

The net head of the power plant was constant and remained 53 m throughout the year, but the power generation depended on user demand and reached a 125 kW peak, as shown in Figure 2. Figure 2 was developed based on details from a household economic status survey, the rural electrification target, and a demand growth forecast. This peak power accounts for the demands of small-scale communal industries, community institutions, and household facilities that are forecasted to be established between five and fifteen years after electrification access is created. This curve helped to refine the original design of the turbine and penstock, which underwent size changes from 150 to 125 kW and from twin penstocks to a single penstock, respectively. The volume of the water varies daily and seasonally because of irrigation activity in the upper and lower streams of the powerhouse, but the water was designed to flow constantly at 0.3 m^3/s by providing a small reservoir, as shown in Figure 1B. The working principle of the RoR design is based on the conversion of the potential (hydraulic) energy of falling water into mechanical energy using a turbine. This turbine is coupled with a generator by a shaft that spins and converts the mechanical energy into electric energy [4–7]. Accordingly, the turbine was selected based on the available head, flow rate [8], construction material, manufacturing capacity, and maintainability [5,9–11].

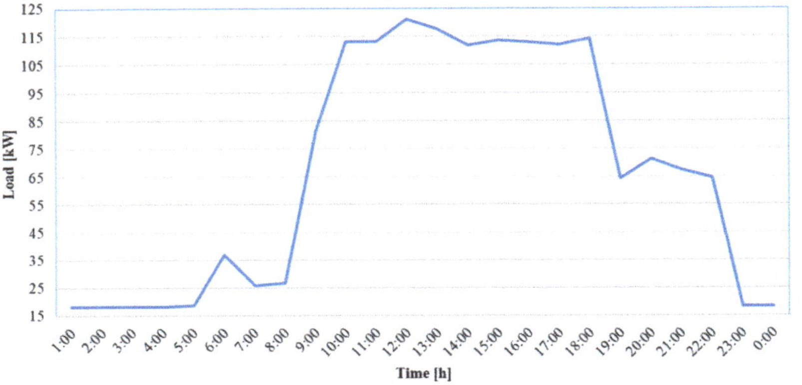

Figure 2. Daily load duration curve of Amentila Village.

1.2. Crossflow Turbine

Crossflow turbines are known for their low costs and ease of design and fabrication and have been extensively adopted in low-head installations. Their performance has been studied extensively through numerical and experimental investigations [9,11–21]. A crossflow turbine is made of a nozzle and a runner. The nozzle guides a water jet into the runner, and the runner converts the energy of the water into mechanical rotation. The runner is composed of several blades that are evenly stacked along the periphery with two or more plates. The nozzle guides the water jet to hit the runner blades at an angle called the angle of attack (α_1). It also has an irregular shape so that a continuous jet of water reaches the runner, transferring the water's energy with minimal losses and the corresponding maximum efficiency [15,20,22–25].

Parameters and Performance

The purpose of this paper is not to present the process of finding the optimal parametric values to ensure the best efficiency of the turbine under consideration but to present the development process of the crossflow turbine to help communities in need to adopt it easily.

Different parameters that affect the performance of crossflow turbines, as indicated by previous studies [4,18–20,22,23,26,27], were considered to optimize the design. These parameters were grouped as operating and geometrical parameters. The operating parameters included the flow rate (Q) and total head (H). The geometric parameters related to the nozzle were the angle of attack (α_1), profile arc (θ), throat width (S_o), etc., and those associated with the runner were the diameter ratio (D_1/D_2), the runner's outer diameter (D_1), the blade thickness (t), the number of blades (n), the size of the shaft, the runner side wall thickness, the alignment of the nozzle, the outer blade's inlet angle (βb_1), the outer blade's exit angle (βb_2), the radius of blade curvature (ρ), and so on. Other parameters, such as the blade thickness, runner side casing thickness, and shaft size, were associated with the structural design, and they were considered in strength design.

In this paper, the 16° angle of attack designed by Banki was adopted. Although some studies claim that maximum efficiency can be achieved with a 20–24° angle of attack, others indicate that efficiency decreases with increases in the angle of attack and suggest using smaller angles [22,24,26–29]. The nozzle arc, with an associated "admittance arc angle", is part of the runner circumference that receives water from the nozzle. In this design, a horizontal nozzle with an admission arc angle of 114° was used to favor installation. Similarly, studies on the number of turbine blades indicated a moderate effect on efficiency [12,13,24]. In addition, the author's blade number optimization study agrees with values in the literature [30]. Increasing the number of blades increases the efficiency of energy transfer but also increases the weight of the rotor and reduces the shaft's output power. Likewise, reduced blade spacing may create back-pressure in the nozzle and may cause efficiency to drop.

2. Methodology

Generally, there are two methods used to design turbines, direct design and inverse design [12,24]. In this work, the direct design method was adopted and ANSYS Computational Fluid Dynamics (CFD) was used to simulate the flow field of the turbine. The direct design method involved two steps: preliminary design, and detailed design and optimization. During preliminary design, the turbine, turbine configuration, and geometric sizing parameters were selected based on site data, including the expected power, head, and flow rate. In the detailed design phase, detailed investigation of the specific flow on the blades and improvement of the design through CFD were performed. In this process, the entire flow mechanism was studied to improve performance by considering the number of blades and the blade profiles. In addition, structural analysis was carried out and simulated to achieve the required strength. This two-step design method was iterated until the optimal result was obtained. A CFD computation flow chart is presented in Figure 3.

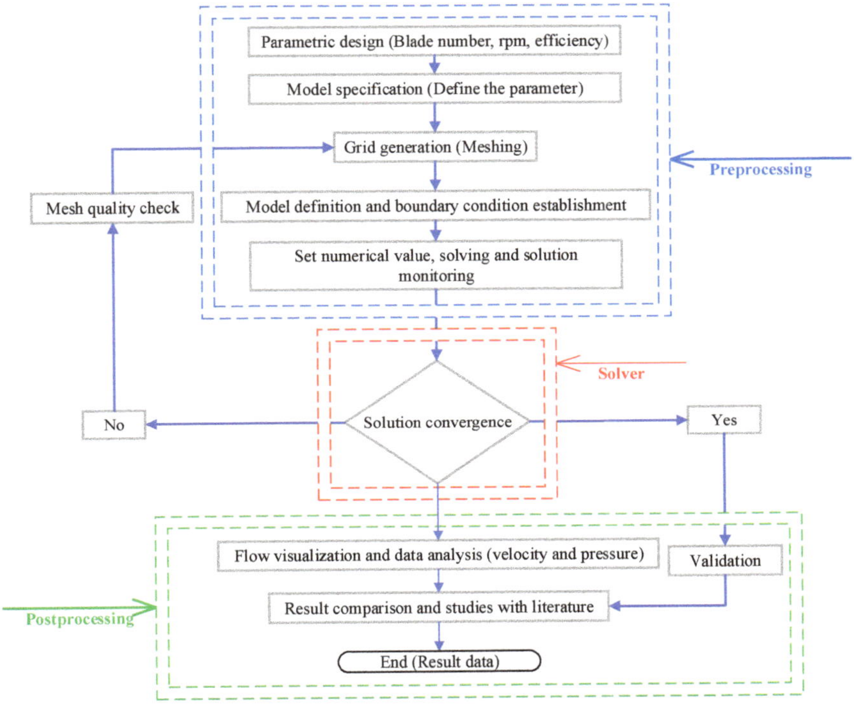

Figure 3. Flow chart of CFD workflow using ANSYS CFX.

Developing the turbine was challenging, as there was no local experience related to turbine manufacturing in the country. Scaled turbine manufacturing was first conducted using Mekelle University's workshop facilities and continued at local flexible manufacturing factories that have modern machining and heat treatment facilities. This process provided practical lessons regarding manufacturing challenges, the requirements of manufacturing facilities, quality control, and following standard production procedures. Later, the full-scale turbine manufacturing documented in this manuscript was contracted to Mesfin Industrial Engineering, a local engineering procurement and manufacturing company, with the authors' supervision and approval of each production step. The turbine was manufactured using numeric control machines, welded by internationally certified welders, and passed the quality control standard of the industry. The technology was simple to transport and was easily assembled and disassembled at the site. A simple manual flow-regulating device was used to adjust the output as per the consumers' load, so this technology could be used by unskilled operators. Through this process, university–industry linkage in solving practical challenges was demonstrated and the technology was localized.

3. Turbine Design

The Amentila hydropower plant has the operating parameters shown in Table 1, and they were used as input data to size the crossflow turbine.

Table 1. Input data.

Parameter	Unit	Symbol	Value
Effective head	m	H	53
Max discharge at effective head	m^3/s	Q	0.3
Frequency	Hz	f	50
Generator speed selection (1000–1500)	rpm	n	1500
Length of penstock	m	Lp	75
Acceleration due to gravity	m/s^2	g	9.81
Water density at 4 °C	kg/m^3	ρ	1000

3.1. Power Calculation

The theoretical power output was calculated using Equation (1). This was the maximum electric power generated by the turbine.

3.2. Geometrical Sizing of Runner

The runner's size was expressed in terms of the runner's length (L) and diameter (D) and the jet thickness (s_o). The width of the rectangular orifice was always equal to the runner's length, while the second cross-sectional dimension, jet thickness, was designed for optimal performance. The data used in the design calculations are shown in Table 2 and the design procedures is given in Appendix A. Consequently, the details of the geometrical sizing of the runner are summarized in Table 3.

Table 2. Design data.

Parameter Description	Parameter	Value
Turbine and generator efficiency	η_e	0.8
Speed ratio	r_t	2
Coefficient accounting for nozzle roughness	c	0.98
Angle of attack	α_1	16°
Inner diameter to outer diameter ratio	i	0.66
Spacing constant	k	0.087
Blade's outer angle	βb_1	30°
Hazen–Williams coefficient	n	0.012

Table 3. The geometrical size of the runner.

No.	Description of Parameter	Parameter	Unit	Equation	Eq.	Value
1	Max turbine output	P_t	kW	$\rho g Q H \eta_e$	(1)	125
2	Runner's outer diameter	D_1	m	$\left(39.81 \times \sqrt{H}\right)/N$	(2)	0.38
3	Runner's rpm	N	rpm	$1500/r_t$	(3)	750
4	Water jet velocity	V_1	m/s	$c \times \sqrt{(2 \times 9.81 \times H)}$	(4)	31.6
5	Runner's tangential velocity	u_1	m/s	$0.5 \times V_1 - cos\alpha_1$	(5)	15.13
6	Runner's length	L	m	$(2.627 \times Q)/(D_1 \times \sqrt{D_2})$	(6)	0.28
7	Runner's inner diameter	D_2	m	$i \times D_1$	(7)	0.25
8	Radial rim width	a	m	$(D_1 - D_2)/2$	(8)	0.065
9	Thickness of water jet	S_o	m	$0.22 \times Q \times (L \times \sqrt{D_2})$	(9)	0.032
10	Spacing of blades	S_1	m	$(k \times D_1/sin\beta_1)$	(10)	0.067
11	Number of blades	n	---	$(\pi \times D_1)/S_1$	(11)	18.05
12	Radius of blade curvature	ρ	m	$\frac{0.32 \times D_1}{2}$	(12)	0.061
13	Distance of jet from center of shaft	Y_1	m	$(0.1986 - 0.945 \times k) \times D_1$	(13)	0.044
14	Distance of jet from inner periphery of runner	Y_2	m	$(0.1314 - 0.945 \times k) \times D_1$	(14)	0.019
15	Inner blade angle	β_{b2}	°	$2 \times \tan \alpha_1$	(15)	90

3.3. CFD Analysis: Optimal Inlet and Outlet Shapes

The schematic and CAD model details of the turbine are shown in Figures 4 and 5, respectively. These pictures were created using SolidWorks version 2018 based on the parameters presented in Table 3. Accordingly, the runner was designed with eighteen blades, an outer diameter of 380 mm, and a speed of 750 rpm. The optimal operating and design parameters are given in Tables 1 and 3, respectively. The commercial CFD software ANSYS CFX version 19.2 was used for fluid flow analysis.

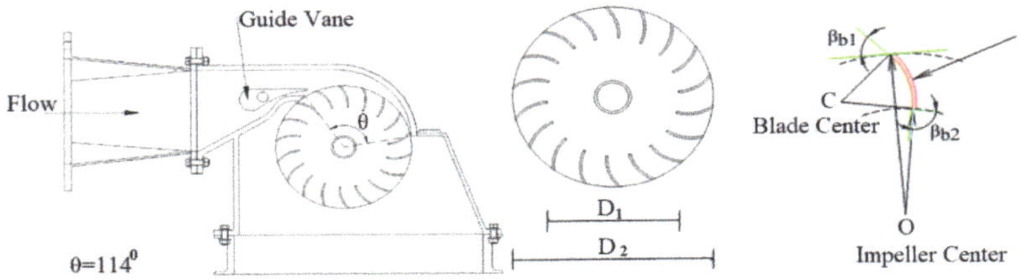

Figure 4. Schematics of turbine nozzle arc, impeller, and blade geometry.

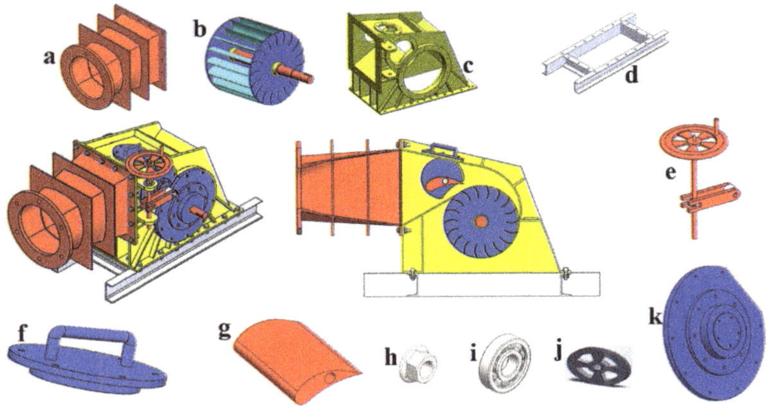

Figure 5. CAD model of turbine components and 3D assembling: (**a**) adapter, (**b**) runner, (**c**) casing, (**d**) skid support, (**e**) power screw, (**f**) top cover, (**g**) guide vane, (**h**) nut, (**i**) bearing, (**j**) wheel, and (**k**) side cover.

3.3.1. Boundary Conditions and Mesh

Boundary conditions were defined according to the geometry of the model shown in Figure 6a. A solid shaft with a diameter of 50 mm was included, so the effect of the angular momentum change in the first stage was considered. The total operating head pressure was specified at the inlet. The inlet was located at the end of the adapter, i.e., between the penstock and the nozzle, and it converted a circular water jet into a rectangular jet, so the flow remained unaffected by any downstream flow irregularities created by the runner's motion. Uniform pressure was assumed at the inlet. The nozzle was considered with a linear inlet velocity at the ambient temperature. At the exit of the turbine, atmospheric pressure was applied. The turbine rotor was considered a solid moving at 750 rpm, and the outer case was analyzed based on fluid flow over a solid part. The guide vane was also cut off from the fluid part. The walls were specified separately with the respective boundary conditions of no-slip, isothermal, stationary walls.

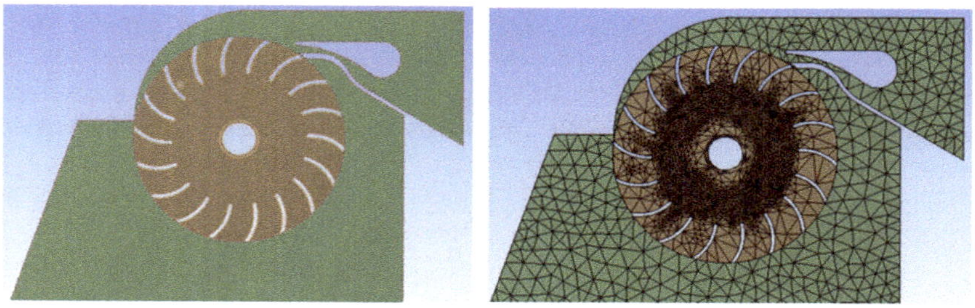

(a) (b)

Figure 6. Structural ANSYS analysis schematics of runner: (**a**) model discretization and (**b**) meshing.

Considering the complex shape of the turbine, an unstructured mesh (prismatic cells) was used to discretize the model. Generation of the computational mesh domain was defined using SolidWorks. The model schematics and mesh (runner and casing) are shown in Figure 6.

3.3.2. Grid Independence Verification

Figure 6a shows the unstructured tetrahedral elements used to discretize the computational flow domain. Unstructured meshing helped to capture precise flow dynamics and automatically adjust the mesh when geometric modifications were made. The boundary surfaces, including the blade, shaft, and casing, were treated using a first-layer inflation technique. Additional layers were added with a 5% growth rate to enhance the resolution of near-wall flow behavior. The wall boundary layer simulation accuracy was assessed using a dimensionless parameter ($Y+$). This parameter gave an estimation of the distance between the wall and the first grid cell, i.e., the thickness of the viscous sublayer. The value of this parameter gave a better representation of the flow separation and the pressure gradient near the walls. The flow profile was validated through mesh refinement analysis. This analysis ensured the simulation results were independent of the discretization parameters. In addition, sensitivity analysis was carried out by selecting torque as a monitoring variable. Accordingly, the result gave an average element size of 3 mm, corresponding to 790,000 mesh elements, and stabilized after 670 iterations.

3.3.3. Momentum and Turbulence (k-ε) Equation

The total pressure was computed based on the dynamic pressure and static pressure of the water flow through the turbine, including the nozzle. Since water is incompressible, the density and temperature of water do not affect the magnitude of pressure. Accordingly, the pressure variation was due to the hydraulic head and the flow of water through different configurations. The dynamic pressure depended on the flow's velocity. This circumstance was easily observed in illustrations of total pressure contours. The static pressure of the flow was computed based on the static head of the flow through the penstock via the nozzle. Since fluid flow properties like density do not change with application of pressure, it was affected by the hydraulic head. In general, the flow was assumed to be isothermal. Temperature variation did not affect the pressure of the water inside the penstock and turbine blades. This analysis helped to check the qualitative and quantitative design results. For practical purposes, emphasis was given to the velocity, flow angle, and pressure, which were found to be significant for turbine performance.

4. Results and Discussion

4.1. Computational Results and Design Parameters

A standard procedure to validate CFD results is to compare the parameters of interest with experimental results. However, no experimental study was carried out. Nevertheless, errors and uncertainties were minimized by iteration convergence and grid independence studies. While iteration convergence helped to ensure converged solutions, grid independence helped to narrow discretization errors. The model was then tested with experimental results from smaller turbines. Afterwards, the model was used to obtain the optimal geometrical shape of the turbine for defined performance flow fields.

The key objective to improve turbine performance was to increase the change in angular momentum at the runner's boundary surface. This was achieved by maximizing the angular momentum of the flow at the inlet and minimizing it at the outlet. From a design point of view, matching the nozzle and impeller geometries was important. By analyzing the flow field around the impeller, it was determined that if the relative angular velocity decreases across a blade row, then work is extracted from the flow, and work is added to the flow when the reverse happens.

Figure 7 shows simulation results for the flow field around the impeller and nozzle. This enabled the pressure, velocity, and flow angles around the impeller to be computed. As can be observed in the velocity profile, the flow at the runner inlet was predominantly in a tangential direction, indicating that the angular momentum flux was expected to increase. However, the flow at the exit of the impeller was in a radial direction, which indicated reduced angular momentum.

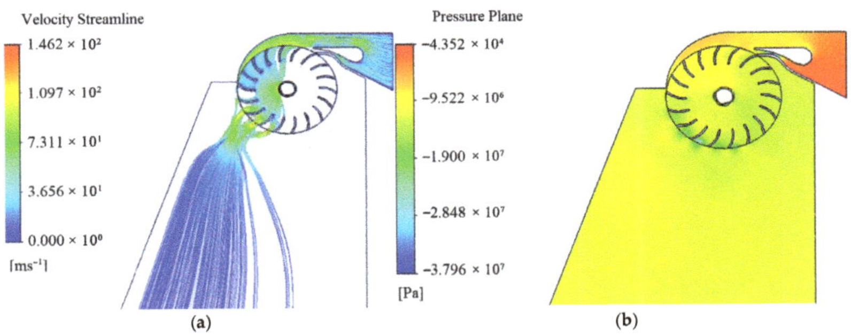

Figure 7. Optimal CFD results: (**a**) velocity profile and (**b**) pressure profile.

4.2. Static Structural Analysis of the Runner Blade

Static structural analysis evaluates stress, deformation, and safety factors of structures subjected to steady static or dynamic loads. This type of analysis is usually used to determine if a structure can safely withstand a specified load by studying the applied stress and yield/ultimate stress of the material used. The von Mises stress criterion is used to predict the yielding of ductile materials. The ratio of yield stress to applied stress is called the safety factor (FS). While an SF < 1 indicates failure risks, an SF > 1 indicates a structure can withstand the applied load and stress. In this paper, a worst-case scenario that was constrained at both ends and a real scenario with a remote displacement constraint were considered to analyze the structural deformation of the runner. The runner was made of structural steel 37 with a yield strength of 235 MPa. The CFD results with maximum deformation and pressure on the blades are presented here for the simulation setup shown in Figure 8. The maximum deformations of the blades at the center were found to be 1.45 mm and 0.14 mm for the first and second scenarios, respectively, as shown in Figure 9a,c. Similarly, the von Mises stress values were found to be 271 MPa and 148 MPa in the first and second scenarios, respectively, as shown in Figure 9b,d. While the first scenario indicated a risk condition with a safety factor of 0.87, the second scenario gave a safe condition with a safety factor of 1.59.

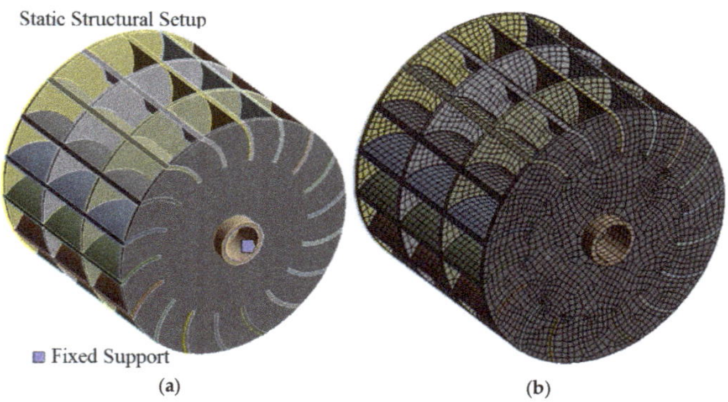

Figure 8. ANSYS setup: (**a**) fixed support setup and (**b**) meshing.

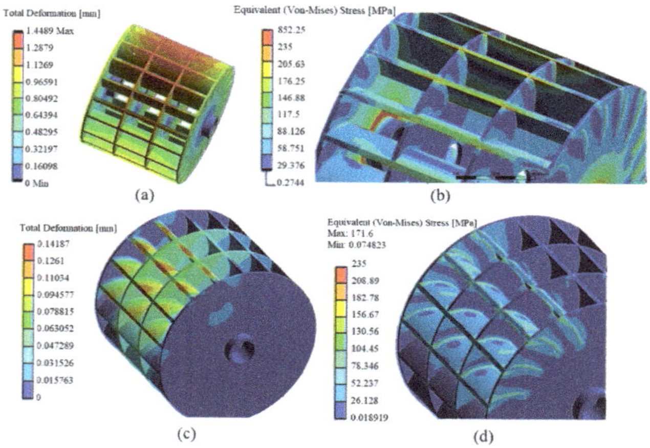

Figure 9. Structural stress behavior: (**a**,**c**) deformation and (**b**,**d**) von Mises stress.

4.3. Modal Analysis of the Runner Blade

Modal analysis is a computational process for predicting modal parameters such as the resonant frequency, mode shape, and damping. It was used during the initial design and design evolution stages that enabled the authors to numerically analyze the turbine model before fabrication. Accordingly, it determined the natural mode shapes and frequencies of the runner during free vibration. Since the designed turbine operates at 750 rpm, its corresponding working frequency was found to be 12.5 Hz using Equation (16).

$$f = \omega / 2\pi \tag{16}$$

Based on this analysis, four first mode shapes named mode 1 to mode 4 with frequencies of 186, 339, 339, and 375 Hz, respectively, were obtained, as shown in Figure 10. Based on this analysis, the working frequency and the natural frequencies of the runner in the four mode shapes were far from each other, indicating the turbine structure has no tendency to resonate.

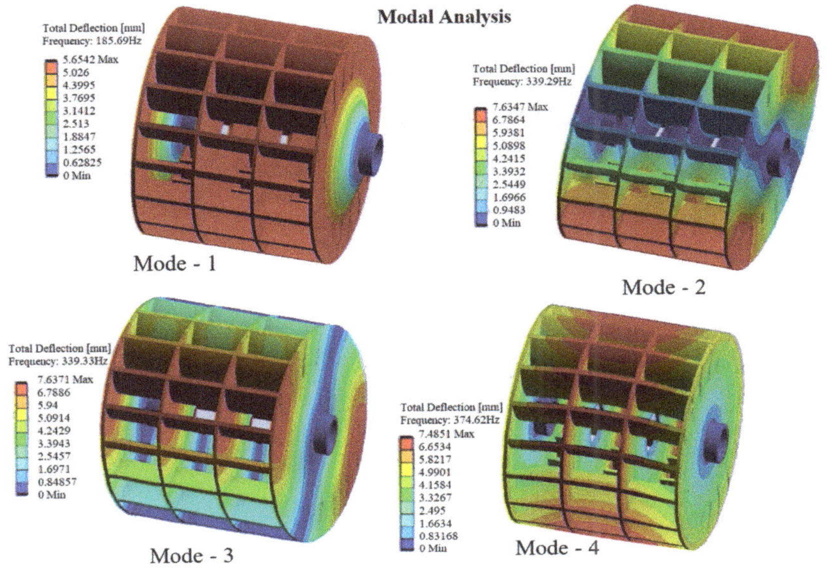

Figure 10. The mode shapes of the runner.

5. Construction

Fabrication of the turbine was performed at Mesfin Industrial Engineering (MIE). This company has no history of developing any kind of turbine but has rich experience in developing hydropower components (steel penstocks, gates, cranes, and other steel structures) for hydropower plants.

The general principle followed here was to avoid complex fabrication techniques like casting, to use locally available materials, and to make assembly and disassembly as simple as possible. Crossflow turbine fabrication involved fabrication of the main parts, such as the support frame, the turbine casing, the runner, a nozzle with an adjustable guide vane, and different-sized bearing holders. The runner was made of three circular discs, a shaft, and eighteen blades, as shown in Figure 11. The three circular discs were grooved into eighteen parts, using a pantograph to hold the blades. Later, they welded on bushes that fit forcefully onto the shaft, as shown in Figure 11b. The shaft was machined in a numeric controlled lathe machine with the required dimensions, surface finish, and tolerance. The support frame was made of standard C-channels welded together, as shown in Figure 11c. The blades were fitted into the grooves of the discs and welded. The assembled runner was then machined to obtain the correct dimensions and mass balance. Two bearings were fitted on the assembled runner before it was fitted in the bearing holders.

Different-sized bearing holders were made for the runner and guide vane assembly, as shown in Figure 12a. These components were first cut in a pantograph with rough dimensions, then refined in a lathe machine for accurate dimensions and tolerances. Different-sized holes were then drilled around these holders to attach them to the casing. Holders with large thicknesses were made by welding two discs. The turbine casing in Figure 13a was made by cutting two side plates, then welding them with support frames. The side plates were first cut by a pantograph, then all the mating parts of these plates were machined for accurate fitting. Later, holes were drilled for standard bolt assembly.

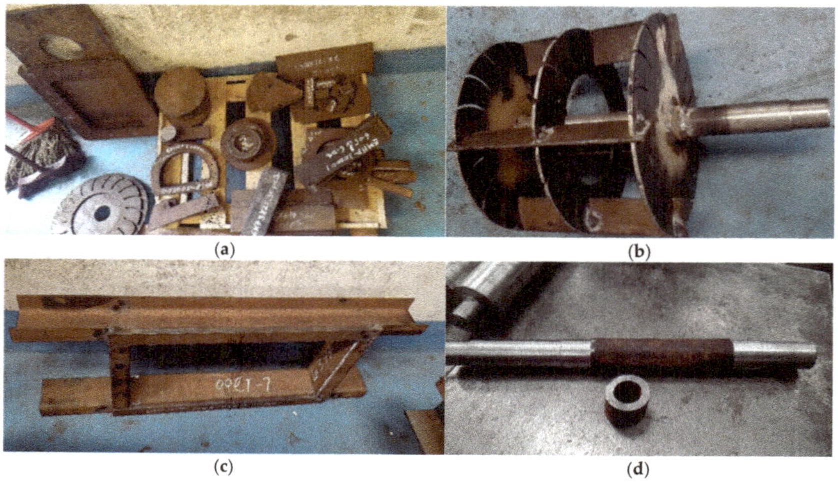

Figure 11. Runner fabrication process: (**a**) component inventory, (**b**) partially assembled runner, (**c**) skid support, and (**d**) shaft and bush.

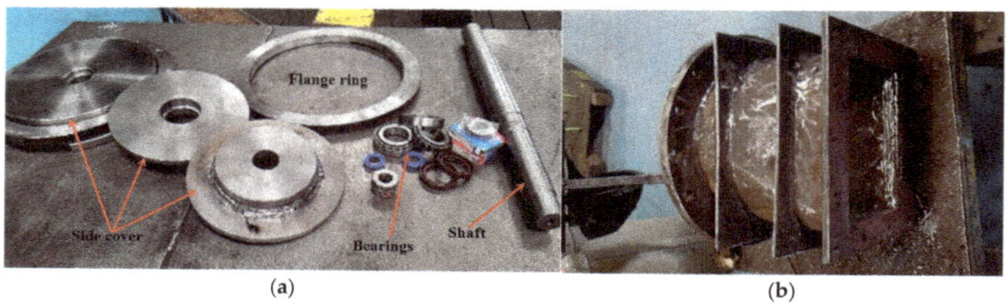

Figure 12. Components of turbine support: (**a**) side cover, flange ring, bearing, and shaft and (**b**) adapter.

With all parts in place, final welding with high strength and no leakage was performed on the casing, as shown in Figure 13c. All bearing holders were then fastened to the casing using bolts. All components were made of steel 37 (St-37), which was imported and easily available in the local market. All bolts, bearings, and flanges used were imported and selected as per the standard.

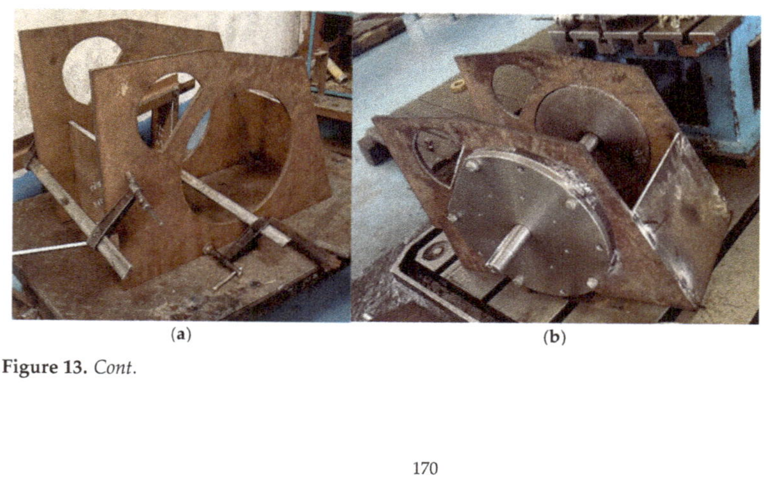

Figure 13. *Cont.*

(c)

Figure 13. Fabrication and assembly process: (**a**) casing arranged for welding, (**b**) shaft and side cover assembly trial, and (**c**) full assembly for inspection.

After quality inspection and testing, all components were painted with three layers of paint, as shown in Figures 14 and 15. The total mass of the assembled turbine in Figure 15 was 6 tons.

Figure 14. Painting and drying components.

Figure 15. *Cont.*

Figure 15. Finishing, receiving, and transportation: (**a**) prepared turbine at manufacturing company, (**b**) transportation to site, and (**c**) prepared turbine at installation site.

6. Conclusions

This research detailed the design, computational analysis, and development of a crossflow turbine. This turbine was engineered based on local hydrological conditions (0.3 m^3/s of discharge and 53 m head) to meet the electricity demand of about 450 households. The turbine has an 18-blade runner with a 16° blade profile, and it was computationally verified to achieve a peak output of 125 kW at 80% efficiency. In addition, the computational results indicated critical performance characteristics, including a constant angle of attack, maximized angular momentum transfer at the runner inlet, and smooth laminar flow in the first and second stages of the turbine. Later the turbine was manufactured, meeting all test standards. Currently, it is ready to be commissioned at the Amentila SHP plant, a standalone off-grid system developed to provide a rural electrification solution. The methodological and technical lessons presented in this paper are important for advancing sustainable off-grid energy solutions in the south. This study serves as a proof of concept for demand-driven off-grid electrification, and it provides essential practical insights for SHP developers and policymakers, highlighting the necessity of designing research strategies based on government policies and changing stakeholders' attitudes to support building local capacity and technology adoption.

Recommendations

Though hydropower turbines have not been designed or developed in Ethiopia and engineering manufacturers are not considering the design and development of turbines as their core business, the authors believe this paper provides unique practical experience that can be considered eye-opening for off-grid electrification stakeholders such as research universities, communities, manufacturing industries, and the power sector (Utility). Accordingly, the authors recommend the following points as a way forward:

i. Hoping existing war crises will improve in the project region, the authors recommend commissioning and performance testing the technology.
ii. Locally manufacturing the crossflow turbine was found to be easy and manageable. The authors observed the SHP project's economic impact on the community, as it promoted irrigation, and the site became a demonstration site where off-grid electrification lessons were given to many rural electrification and rural development stakeholders. The authors encourage these actors to rethink and modify their off-grid electrification and rural development strategies towards effective approaches based on the water–energy–food–environment nexus.

iii. The authors encourage Mekelle University and its stakeholders to push forward and use this intervention as a benchmark to mobilize local communities to cooperate and own such technological solutions through community-owned business models.

Author Contributions: A.H.T. contributed to the conceptualization, method, analysis, interpretation, supervision, drafting and redrafting of the manuscript, and editing. S.A.W. contributed to the method, analysis, interpretation, and drafting of the manuscript. K.G.G. contributed to the method, analysis, and drafting of the manuscript. All authors have read and agreed to the published version of the manuscript.

Funding: This research received no external funding.

Data Availability Statement: Necessary data are contained within this article. However, if further data are required, they will be available on request.

Acknowledgments: The authors want to acknowledge Mekelle University for committing to finance the development of a small hydropower plant to demonstrate water–energy–food–environment nexus research. In addition, the authors wish to express special gratitude to Nigus Abebe, who worked as a civil engineering team leader; Melkamu Nibret, who worked as an electrical team leader; and Henok Mezgebe, who voluntarily supported the mechanical team by providing a licensed computational facility during the design stage. Finally, the authors want to acknowledge Endu Atsbeha for his transportation service in all stages of this work and Zekarias, chief engineer from the manufacturer, for his professionalism and his extraordinary commitment to visit the project site before installation.

Conflicts of Interest: The authors declare no conflicts of interest.

Nomenclature

Symbol/Abbreviation	Description
CFD	Computational Fluid Dynamics
D	Diameter
D1	Runner's outer diameter
D2	Runner's inner diameter
H	Total head
L	Runner's length
n	Number of blades
P	Output power
Q	Flow rate
RoR	Run-of-river
S_o	Throat width/jet thickness
t	Blade thickness
α_1	Angle of attack
βb_1	Outer blade's inlet angle
βb_2	Blade's exit angle
ρ	Radius of blade curvature
θ	Profile arc

Appendix A

Effective head	48.5
Discharge	0.7921365
Power	282.665

No	Turbine type	Selection Criteria									
		Type	Head(m)	Discharge(m3/se)	Power(KW)	Efficiency range	Shaft alignment	Flow direction	Maintainability	Manufacturability	Cost
1	Pelton	Impulse	15-500	0.1-3	100-5000	0.80-0.85	V and H(small and medium)	Tangential		Difficult	60-100
2	Francis	Reaction	15-300	0.3-20	100-5000	0.88-0.92	V and H			Imported	60-120
3	Kaplan		5-80	>0.9						Complicated	179,299.7 pound
4	Propeller		<80			0.88-0.92				Imported	60-120
5	Cross flow	Impulse	5-100	0.1-10	50-1000	0.8	V and H	Tangential	Easy	Easy	50-80
6	Reverse pump		6~100	0.02-0.9	1-300					Mass	cheap
7	Turgo	Impulse	20-250	0.2-8	100-10,000					Imported	80-100
8	Submerged pump type		2.5~20	0.4~10	30-850					Unfit	
9	S shape		3~18	1.5~40	50-5000					Unfit	
10	Vertical shaft Tubular		5~18	2~20	100-2000					Unfit	
11	Bulb(Tubular)		3~18	3~40	100-5000					Unfit	
12	Package Bulb		4~20	4~25	150-3500					Unfit	
13	Straight flow		3~20	0.5~5	10-600					Unfit	

$$P_t = 9.8 \times Q_{max} \times H_e \times \eta_t$$

where,

P_t Maximum turbine output at the effective head H (kW)

Q_{max} : Maximum discharge at the effective head H (m^3/sec)

He : Effective Head (m)

η_t : Water turbine efficiency

$$\eta_t = P_t / (9.8 \times Q_{max} \times H_e).$$

where,

Pt Turbine output (kW)

Qmax Discharge (m^3/sec)

He Effective head (m)

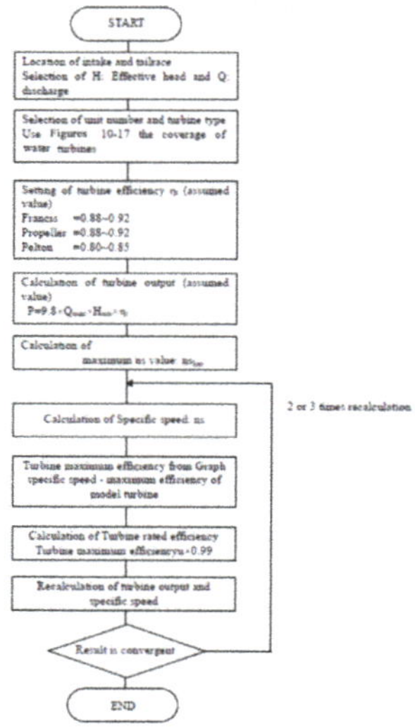

Figure A1. Turbine selection procedure chart.

Turbine type	Cross flow		
Input data			
Description	**Symbol**	**Value**	**Units**
Effective head	H	53	m
Max Discharge at effective head	Q	0.3	m3/s
Frequency	f	50	Hz
Turbine efficiency setting(assumption)	η	0.8	
generator rpm selection(1000-1500)	N	1500	rpm
speed ratio	r	2	
coeffient counting for nozzle roughness	C	0.98	
Angle of attack	α1	16	
inner to ouer diameter ratio		0.66	
Spacing constant (0.075-0.1)	k	0.087	
Blade inlet angle sin β1(30)	β1	0.5	
Hazen Williams coefficient	n	0.012	
Length of penstock	L	75	m
Runner design			
Description	**Symbol**	**Value**	**Units**
Turbine efficiency setting(assumption)	ηt	0.8	
Max Water turbine output at effective head	Pt	124783.2	W
Runner outer Diameter	D1	0.386428233	m
runner rpm	N	750	rpm
Water jet velocity(runner tangential velocity)	V1	31.60192311	m/s
Runner tangential velocity/tangential velocity of runner outer periphery	u1	15.13194063	m/s
length of the turbine runner	L	0.280139625	m
Runner inner diameter	D2	0.255042634	m
radial rim width	a	0.0656928	m
thikness of water jet	So	0.03236171	m
spacing of blades	S1	0.067238513	m
number of blades	n	18.04597701	
radius of blade curvature	ρ	0.061828517	m
Distance of jet from center of shaft	Y1	0.04497445	m
Distance of jet from inner periphery of the runner	Y2	0.019006473	m
Angle b/n relative velocity of entering water jet & outer runner periphery	β1	0.601264484	
Penstock Design			
Description	**Symbol**	**Value**	**Units**
Penstock diameter	D	**0.342409072**	m
Length of pipeline	Lp	75	m
Pipe area	A	0.092036518	sqm
Friction factor according to moody diagram	λ	0.036	
Pipe velocity	v	3.259575715	m/s
Head loss-due friction	hfriction	1.897837954	m
Absolute roughness k in mm	k	0.0015	mm
relative roughness	k/D	4.38073E-06	
Rainolds number	Re	1116108.295	
loss coefficients	0.25	inlet	
	0.04	Reducer	
	0.245	Elbow(<45)	
	0.245	Elbow(<45)	
	0.044	Elbow(<15)	
	0.044	Elbow(<15)	
	0.86	Valve	
	0.22	Regulator(BF)	
Other head losses	1.05490195		m
Head loss	%	5.6%	Percent
For power transmission and shaft design			
Torque	T	1589.594904	
	Fr	w	
	Ft	8227.11577	
	Wt	4113.557885	
	Wr	10080.22226	

Figure A2. Crossflow turbine design procedure.

Parameter	Symbol	
Torque	T	1589.59
Tangential force	Ft	8227.12
Half Tangential force	Wt	4113.56
Half Radial force	Wr	10,080.22
Pulley tension	Tp	20,000.00
Length	L1	0.01
	L2	0.65
	L3	0.01
	L4	0.25
Vertical reaction	RVA	10,080.22
	RVD	9929.77
Horizontal reaction	RHA	−3349.13
	RHD	31,576.24
Horizontal Bending Moment	MHA	0.00
	MHB	−33.49
	MHC	−4884.24
	MHD	−5000.00
	MHE	0.00
Vertical Bending Moment	MVA	0.00
	MVB	100.80
	MVC	100.80
	MVD	0.00
	MVE	0.00
Resultant Bending Moment	MRA	0.00
	MRB	106.22
	MRC	4885.28
	MRD	5000.00
	MRE	0.00

From pulley design

	Allowable stress in shear (MPa)		54.0
Location	Torque	Eq Twisting moment	Shaft diameter(mm)
A	0.0	0.0	0.000
B	1589.6	1593.1	53.172
C	1589.6	5137.4	78.556
D	1589.6	5246.6	79.109
E	1589.6	1589.6	53.133

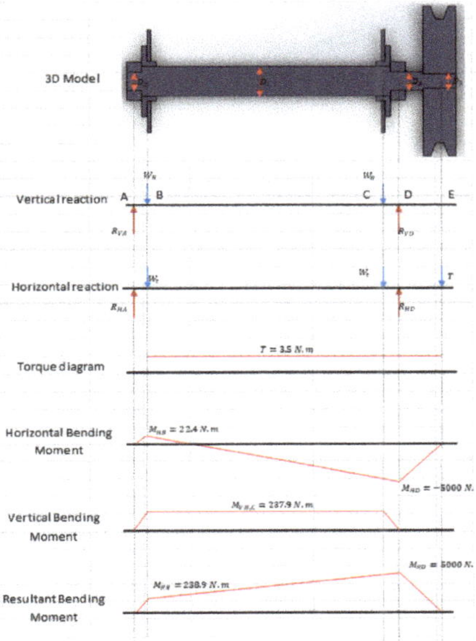

Figure A3. Shaft design.

References

1. Wu, G.C.; Deshmukh, R.; Ndhlukula, K.; Radojicic, T.; Reilly, J. *Renewable Energy Zones for the Africa Clean Energy Corridor*; Technical Report Number: LBNL-187271; Lawrence Berkeley National Laboratory: Berkeley, CA, USA, 2015. [CrossRef]
2. Dalelo, A. Rural Electrification in Ethiopia: Opportunities and Bottlenecks. Ph.D. Thesis, Addis Ababa University, Addis Ababa, Ethiopia, 2003. Available online: https://www.academia.edu/115469209/Rural_Electrification_in_Ethiopia_Opportunities_and_Bottlenecks (accessed on 29 June 2025).
3. Cheever, M.; Graichen, K.; Homeier, D.; Howell, J.; Kefauver, O.; Kimball, T. Environmental Policy Review: Key Issues in Ethiopia 2011. In *Chapter 6 Rural Electrification & Renewable Energy in Ethiopia*; Colby College Environmental Policy Group: Waterville, Maine, 2011; pp. 1–29. Available online: https://web.colby.edu/eastafricaupdate/ (accessed on 29 June 2025).
4. Elbatran, A.; Yaakob, O.; Ahmed, Y.M.; Shabara, H. Operation, performance and economic analysis of low head micro-hydropower turbines for rural and remote areas: A review. *Renew. Sustain. Energy Rev.* **2015**, *43*, 40–50. [CrossRef]
5. Razak, J.A.; Ali, Y.; Alghoul, M.; Zainol, M.S.; Zaharim, A.; Sopian, K. Application of Crossflow Turbine in Off-Grid Pico Hydro Renewable Energy System. *Proceeding Am.-Math* **2010**, *10*, 519–526.
6. Du, J.; Yang, H.; Shen, Z.; Guo, X. Development of an inline vertical cross-flow turbine for hydropower harvesting in urban water supply pipes. *Renew. Energy* **2018**, *127*, 386–397. [CrossRef]
7. Achebe, C.H.; Okafor, O.C.; Obika, E.N. Design and implementation of a cross fl ow turbine for Pico hydropower electricity generation. *Heliyon* **2020**, *6*, 04523. [CrossRef] [PubMed]
8. McLean, E.; Kearney, D. An Evaluation of Seawater Pumped Hydro Storage for Regulating the Export of Renewable Energy to the National Grid. *Energy Procedia* **2014**, *46*, 152–160. [CrossRef]
9. Vortex, F.; Sritram, P.; Suntivarakorn, R. The Efficiency Comparison of Hydro Turbines for Micro Power. *Energies* **2021**, *14*, 7961. [CrossRef]
10. Adhikari, R.; Wood, D. Computational Analysis of a Double-Nozzle Crossflow Hydroturbine. *Energies* **2018**, *11*, 3380. [CrossRef]
11. Dragomirescu, A.; Schiaua, M. Experimental and Numerical Investigation of a Bánki Turbine Operating far away from Design Point. *Energy Procedia* **2017**, *112*, 43–50. [CrossRef]

12. Adhikari, R. Design Improvement of Crossflow Hydro Turbine. Ph.D. Thesis, University of Calgary, Calgary, AB, Canada, 2016. [CrossRef]
13. Adhikari, R.; Wood, D. The Design of High Efficiency Crossflow Hydro Turbines: A Review and Extension. *Energies* **2018**, *11*, 267. [CrossRef]
14. Sinagra, M.; Tucciarelli, T.; Picone, C.; Aricò, C.; Hannachi, M. Design of Reliable and Efficient Banki-Type, Environ. *Sci. Proc.* **2020**, *2*, 49. [CrossRef]
15. De Andrade, J.; Curiel, C.; Kenyery, F.; Aguillón, O.; Vásquez, A.; Asuaje, M. Numerical Investigation of the Internal Flow in a Banki Turbine. *Int. J. Rotating Mach.* **2011**, *4*, 14. [CrossRef]
16. Mehr, G.; Durali, M.; Khakrand, M.H.; Hoghooghi, H. A novel design and performance optimization methodology for hydraulic Cross-Flow turbines using successive numerical simulations. *Renew. Energy* **2021**, *169*, 1402–1421. [CrossRef]
17. Sinagra, M.; Picone, C.; Aricò, C.; Pantano, A.; Tucciarelli, T.; Hannachi, M.; Driss, Z. Impeller Optimization in Crossflow Hydraulic Turbines. *Water* **2021**, *13*, 313. [CrossRef]
18. Totapally, H.G.; Aziz, N.M. Refinement of Cross-Flow Turbine Design Parameters. *J. Energy Eng.* **1994**, *120*, 133–147. [CrossRef]
19. Budiarso, W.; Adanta, D.; Iaeng, M. Computational analysis of flow field on cross-flow hydro turbines. *Eng. Lett.* **2021**, *29*, 87–94.
20. Sammartano, V.; Aricò, C.; Carravetta, A.; Fecarotta, O.; Tucciarelli, T. Banki-Michell Optimal Design by Computational Fluid Dynamics Testing and Hydrodynamic Analysis. *Energies* **2013**, *6*, 2362–2385. [CrossRef]
21. Sammartano, V.; Morreale, G.; Sinagra, M.; Tucciarelli, T. Numerical and experimental investigation of a cross-flow water turbine. *J. Hydraul. Res.* **2016**, *54*, 321–331. [CrossRef]
22. Choi, Y.D.; Lim, J.I.; Kim, Y.T.; Lee, Y.H. Performance and Internal Flow Characteristics of a Cross-Flow Hydro Turbine by the Shapes of Nozzle and Runner blade. *J. Fluid Sci. Technol.* **2008**, *3*, 398–409. [CrossRef]
23. Mrope, H.A.; Abeid, Y.; Jande, C.; Kivevele, T.T. A Review on Computational Fluid Dynamics Applications in the Design and Optimization of Crossflow Hydro Turbines. *J. Renew. Energy* **2021**, *2021*, 7048. [CrossRef]
24. Adhikari, R.; Wood, D. A new nozzle design methodology for high efficiency crossflow hydro turbines. *Energy Sustain. Dev.* **2017**, *41*, 139–148. [CrossRef]
25. Warjito, W.; Budiarso, B.; Celine, K.; Nasution, S.B.S. Computational Method for Designing a Nozzle Shape to Improve the Performance of Pico-Hydro Crossflow Turbines. *Int. J. Technol.* **2021**, *12*, 139–148. [CrossRef]
26. Chen, Z.M.; Choi, Y.D. Performance and internal flow characteristics of a cross-flow turbine by guide vane angle. *IOP Conf. Ser. Mater. Sci. Eng.* **2013**, *52*, 052031. [CrossRef]
27. Du, J.; Shen, Z.; Yang, H. Effects of different block designs on the performance of inline cross-fl ow turbines in urban water mains. *Appl. Energy* **2018**, *228*, 97–107. [CrossRef]
28. Adanta, D.; Warjito, B.; Prakoso, A.P.; Wijaya, E.P. Effect of Blade Depth on the Energy Conversion Process in Crossflow Turbines. *CFD Lett.* **2020**, *12*, 123–131.
29. Bachant, P.; Wosnik, M. Effects of Reynolds Number on the Energy Conversion and Near-Wake Dynamics of a High Solidity Vertical-Axis Cross-Flow Turbine. *Energies* **2016**, *9*, 73. [CrossRef]
30. Assefa, E.Y.; Tesfay, A.H. Effect of Blade Number on Internal Flow and Performance Characteristics in Low-Head Cross-Flow Turbines. *Energies* **2025**, *18*, 318. [CrossRef]

MDPI AG
Grosspeteranlage 5
4052 Basel
Switzerland
Tel.: +41 61 683 77 34

Energies Editorial Office
E-mail: energies@mdpi.com
www.mdpi.com/journal/energies

www.ingramcontent.com/pod-product-compliance
Lightning Source LLC
Chambersburg PA
CBHW061605210926
51850CB00009B/2215

* 9 7 8 3 7 2 5 8 5 8 0 9 5 *